POLYMER SEQUENCE DETERMINATION

Carbon-13 NMR Method

POLYMER SEQUENCE DETERMINATION
Carbon-13 NMR Method

JAMES C. RANDALL
Research and Development
Phillips Petroleum Company
Bartlesville, Oklahoma

1977

ACADEMIC PRESS New York San Francisco London

A Subsidiary of Harcourt Brace Jovanovich, Publishers

ACADEMIC PRESS, INC.
111 Fifth Avenue, New York, New York 10003

United Kingdom Edition published by
ACADEMIC PRESS, INC. (LONDON) LTD.
24/28 Oval Road, London NW1 7DX

Library of Congress Cataloging in Publication Data

Randall, James C.
Polymer sequence determination.

Includes bibliographical references.
1. Polymers and polymerization—Spectra.
2. Nuclear magnetic resonance spectroscopy. 3. Carbon
—Isotopes—Spectra. I. Title.
QC463.P5R36 541'.2254 77-6610
ISBN 0-12-578050-8

PRINTED IN THE UNITED STATES OF AMERICA

Contents

3 Number-Average Sequence Lengths in Copolymers and Terpolymers

4 Statistical Analyses of Monomer Distributions and Number-Average Sequence Lengths

5 Experimental Design for Quantitative FT-NMR Measurements

6 A Survey of Carbon-13 NMR Studies of Vinyl Homopolymers and Copolymers

Preface

With the development of polymer structural characterizations using NMR, there has been considerable effort expended in measurements of tacticity, sequence distributions, and number-average sequence lengths. A central source of information is needed to stimulate the use of these recently developed methods for polymer characterization, particularly since most of the newer methods utilize ^{13}C NMR. This book was written to provide a background in the methods and applications of ^{13}C NMR to determine polymer structure. It is designed primarily as a reference for those polymer scientists interested in structural characterizations, but it may also benefit NMR spectroscopists who are interested in polymer studies. To avoid mathematical and phenomenological descriptions well covered in more basic texts, the NMR discussions are restricted to spectral interpretations, applications, and experimental considerations. The overall scope is to place ^{13}C NMR polymer structure determinations into a framework understood by the polymer scientist.

Chapter 1 offers a brief review of polymer structure with emphasis placed upon those structural features delineated by ^{13}C NMR. Also included is a detailed discussion of assignment techniques used during ^{13}C NMR interpretations of polymer spectra. In Chapters 2 and 3, methods are presented for measuring sequence distributions and number-average sequence lengths for configurational sequences in vinyl homopolymers and for the comonomer distribution in copolymers and terpolymers. In particular, it is shown how comonomer sequence distributions are related to number-average sequence lengths. This information is usually supplementary; however, there are conceptual advantages to expressing compositional data in each form. Chapter 4 contains a discussion of statistical approaches to polymer characterization. This chapter provides an introduction to statistical analyses and should prepare the reader for more advanced or in-depth studies. Chapter 5 contains practical experimental considerations that one must face when designing an NMR experiment to obtain quantitative structural information. Finally, Chapter 6 presents a critical review of ^{13}C NMR studies reported in the literature for various vinyl homopolymers and copolymers. The spectra presented were obtained in the author's laboratory so the data are for the same magnetic field strength and comparable experimental conditions. Hexamethylsilane was used as the internal standard and the chemical shifts were corrected to the more common tetramethylsilane (TMS) internal standard. Although various literature studies are presented in detail, many of the examples given in the initial phases of the text are from studies conducted in the author's laboratory. This approach was selected to provide the reader with relatively easy examples of the type of structural information available from ^{13}C NMR.

Acknowledgments

The author gratefully acknowledges the assistance of Gerard Kraus, Paul R. Stapp, and John R. Paxson who read preliminary versions of the manuscript and offered worthwhile improvements and suggestions. The author is also indebted to Fred L. Tilley for obtaining the ^{13}C NMR spectra shown throughout the text. Finally, appreciation is expressed to the many colleagues who took time for discussions and to the Phillips Petroleum Company for providing the facilities used in preparing the manuscript and figures.

POLYMER SEQUENCE DETERMINATION

Carbon-13 NMR Method

1 Polymer Structure and Carbon-13 NMR

A major accomplishment of ^{13}C NMR studies of addition polymers has been the bridging of the gap between actual structure and those structural features discernible to the polymer scientist. A brief review of the structural variabilities possible among addition polymers, therefore, is necessary before beginning a discussion about ^{13}C NMR. Generally, polymers prepared by addition polymerizations are necessarily characterized by oversimplified representations. Monomers may add to a growing polymer chain in more than one manner; for some polymers this is no more than a difference in monomer unit configuration, for others it may be a skeletal rearrangement that produces both branched and linear units, as in polybutadienes. Only average structures, however, can be determined. These may not reflect the sequencing of monomer units or the structural variabilities possible from one polymer chain to another.

A crucial problem facing the polymer chemist is to determine structure as accurately and descriptively as possible. A true structural determination would describe the monomer sequences or sequence distributions for each chain length. As stated previously, final averages cannot reflect these structural characteristics. More detailed average structures, however, that describe average sequence distributions and number-average sequence lengths of monomer additions have now become available.

Average polymer structural information, expressed in its simplest form, is given by the monomer distribution, that is, the ratio of monomer A to monomer B in copolymers, or the percent of monomer as structural entity 1, 2, etc., in homopolymers. Monomer distributions of this type can be measured in a variety of ways: structural degradation to produce individual monomer units for quantitative analyses, and spectroscopic determinations using either infrared, ultraviolet, or NMR techniques. We should be reminded, however, that simple monomer distributions reveal no information concerning the sequencing of polymer chain units or the distribution of structures that are so important in determining polymer physical properties.

Sequencing information can be obtained for copolymers if the monomer distribution is expressed in terms of combinations of adjacent structural units. For example, a distribution of AA versus AB versus BB allows number-average sequence lengths to be determined in copolymers. Until recently, such information was not readily available. For the first time, specifically through ^{13}C NMR, concentrations of unlike connecting units in copolymers can be measured independently of concentrations of like connecting units. This result leads to complete sequence distributions and to number-average sequence lengths for monomer additions of the same type. In an early review, Farrar (1) stated that "^{13}C NMR spectroscopy is probably the single most important tool for determining the structure of organic and biological compounds and, as such, it is almost impossible to overestimate its importance." This statement applies equally to structure determinations of polymers.

Since ^{13}C NMR will be the only source of polymer structural information used throughout this book, some advantages offered by magnetic resonance methods will be discussed briefly. One fact not often mentioned in polymer structural determinations is that an NMR spectrum represents directly and completely the "average molecule." No extinction coefficients are required and each resonance area is directly proportional to the number of contributing species.† The NMR spectrum, as recorded, is a result of accumulations of responses from millions of individual polymer molecules. Thus we see and interpret results from a final summation that represents directly the average polymer chain. The principal advantage of ^{13}C NMR, however,

† Initially, there was concern about possible differences in nuclear Overhauser effects (NOE) for polymers in solution NMR studies. (See Section 5.2.) However, NOE differences are not expected theoretically and none have been observed for polymers in solution; consequently, the NOE is not a consideration in quantitative ^{13}C studies of most polymers.

is a sensitivity toward subtle structural features displayed in an accessible form.

Chemical shift differences are observed in ^{13}C polymer spectra for carbon skeleton rearrangements of repeat units, inversions in head-to-tail monomer additions, and for configurational sequences from two to seven units in length. An interpretation of ^{13}C NMR polymer spectra can, therefore, be a perplexing analytical problem. Spectral responses associated with stereochemical configurations are often superimposed upon a corresponding response to the polymer carbon skeleton or to the differences that occur because of changes in the modes of monomer additions. When describing polymer structure, one customarily divides the polymer backbone conceptually into a succession of individual monomer units, even though after a polymer has formed, the source of any particular backbone carbon may be arbitrary. One must be able to relate observed, complex ^{13}C NMR spectral patterns to successions of monomer additions. This translation of an observed ^{13}C NMR polymer spectrum to a proposed final average structure is the topic to which this book is principally concerned.

In subsequent discussions of ^{13}C NMR of polymers, we shall use the NMR spectrum as a starting point for determinations of structures of average polymer chains. Because the NMR phenomenon has been well described previously, experimental development of ^{13}C NMR quantitative methods for describing polymer structure will be emphasized and only a casual acquaintance with the NMR phenomenon will be required of the reader. The NMR discussions will essentially involve ^{13}C chemical shift behavior and assignments in polymer spectra and the methods for obtaining number-average sequence lengths from NMR data. For a more fundamental NMR background and the basic Fourier transform concepts, the reader is referred to Farrar and Becker (2).

The remainder of this introductory chapter contains a discussion of ^{13}C NMR assignment techniques and an interpretation of ^{13}C chemical shift behavior in vinyl homopolymers. We shall discuss polymer structure in detail, and at the risk of being considered artificial, polymer structure will be viewed as a succession of individual monomer units whether configurational relationships or comonomer distributions are being defined. For consistency and simplicity, polypropylene has been chosen as an example of a typical vinyl homopolymer throughout the first two chapters. The concepts introduced, however, apply to any polymeric system as will be seen in this and later chapters. With assignments established, the remainder of the text is devoted to the use of ^{13}C NMR data for quantitative structural determinations in polymers, that is, tacticity measurements and determinations of comonomer distributions and number-average sequence lengths.

1.1 CONFIGURATIONAL SEQUENCES IN VINYL HOMOPOLYMERS

Before beginning our discussion of the ^{13}C NMR chemical shift behavior in polymers, let us examine some of the more general structural features found in vinyl homopolymers,

$$\left[\begin{array}{c} CH-CH_2 \\ | \\ R \end{array} \right]_n$$

where R can be OH, CN, Cl, acetate, methyl, phenyl, etc. These polymers consist of a series of alternating methylene and methine carbons where the latter can exist in a configuration that is either like or unlike that of the immediately preceding methine carbon. In his original definition of polymer configurational relationships, Natta (3) used *isotactic* to describe successive monomer units which, upon appropriate translation and rotation, were configurationally superimposable. The term *syndiotactic* was introduced to describe successive monomer units where inversion of configuration occurred, and any rotation or translation of either unit would not result in superposition. Although these definitions are still used as Natta intended, the term *isotactic* now applies to any number of successive monomer units with the same relative configuration, and *syndiotactic* to any number of alternating configurations.

Differences in configuration of successive monomer units lead to unique relative configurations but not to absolute configurations that can be determined, thus *d,l* stereochemical designations are not used. Only relative configurations can be designated, that is, as either *like* or *opposite*. In the strictest sense, vinyl polymer methine carbons are asymmetric; however, two of the groups attached to the asymmetric carbon become essentially equivalent for long chains. It is only the methine carbons which are close to the terminal units that have sufficiently different structural groups attached to be considered truly asymmetric. Within interior sequences, the configurational *sense* is retained from one unit to the next, although any optical activity is *lost*. Conceptually, this point may be realized through the following projections:

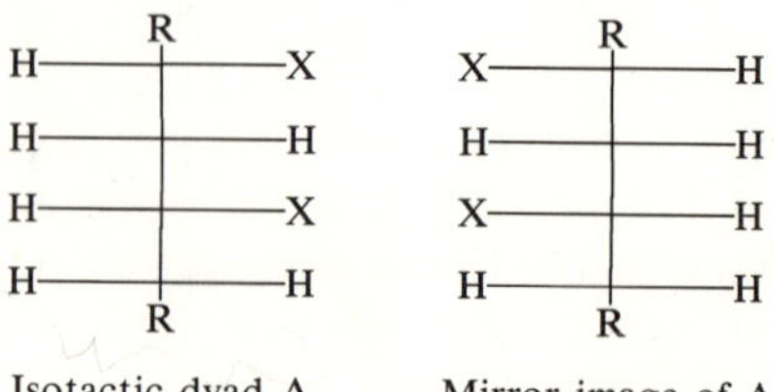

Isotactic dyad A Mirror image of A

If the R groups are of sufficient length to be considered equivalent, a rotation of either unit 180° in the plane of the paper results in images that differ only by the sequence of the four carbon atoms between the R groups. Although one may argue that the mirror images above are not strictly superimposable, translation of one-half of a monomer unit from an R group will produce superposition as shown in Fig. 1.1. Thus, even if interior sequences of adjacent vinyl polymer chains could be compared, one could not distinguish absolute configurations. For a practical standpoint, the methine carbons in vinyl polymers are frequently called *pseudoasymmetric* (4, 5) rather than *asymmetric*.

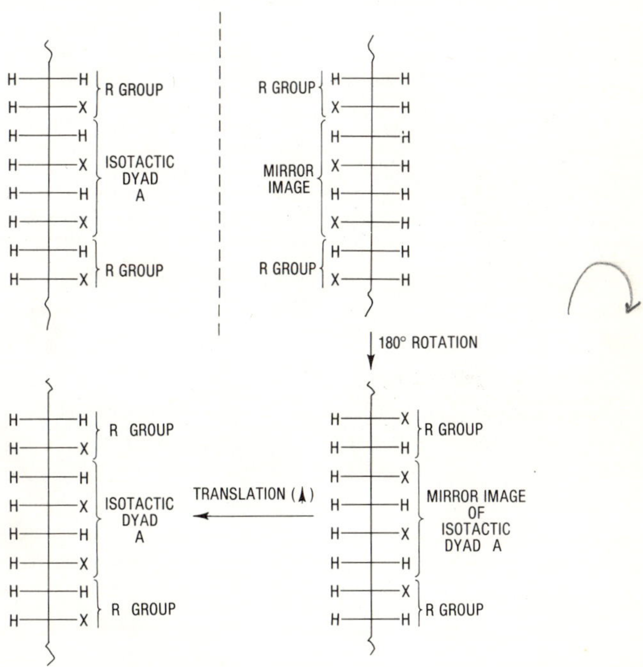

Fig. 1.1. Equivalence of dyads of opposite "handedness" in adjacent chains.

Configurational differences within a polymer chain, of course, do exist or we would not have syndiotactic sequences. Relative configurations among a succession of monomer units can be discriminated and isotactic placements of opposite handedness or chirality can exist within a single chain. For example, successive dyad placements with an opposite handedness appear as follows:

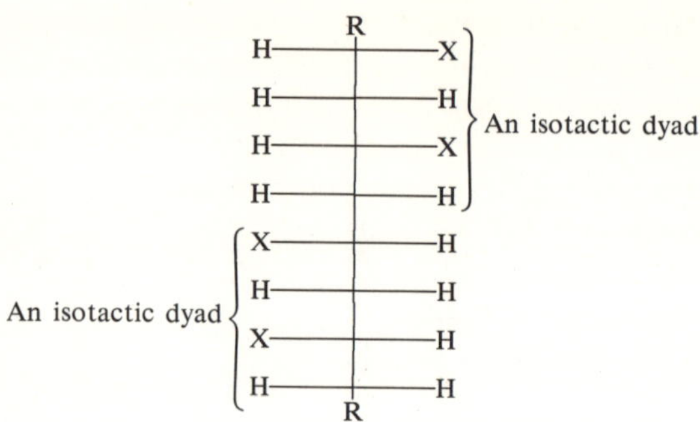

Although opposite isotactic dyad placements are depicted above, specific absolute configurations cannot be assigned because, once again, a 180° rotation of the mirror image and subsequent translation would produce superposition. This failure to detect absolute configurations occurs in polymers because two of the attached groups are sufficiently long to be considered equivalent.

To account for differences in monomer unit chirality, Price (6) introduced a *0* and *1* nomenclature to depict individual monomer unit configurations along a polymer chain. A *0* is used to designate any particular configuration and a *1* denotes its counterpart. Since an inference of absolute configuration must be avoided, the 0's and 1's can be interchanged for any particular sequence. For example, the above projection may be specified by either 0011 or 1100 and the 0's and 1's are used strictly to identify like and unlike configurations. Relative configurational differences are best described by a system developed by Bovey (4, 7–9) which elegantly accounts for configurational differences without inferring chirality. Adjacent monomer pairs with the same relative methine configurations are called *meso* (*m*) dyads while those with opposite configurations are called *racemic* (*r*) dyads. Thus *mrm* applies equally to the 0011 or the 1100 sequence.

We may question the wisdom of introducing two different systems of nomenclature when one should suffice in describing polymer structure. As it will be seen later, there are conceptual advantages to each. An inference of chirality is avoided with the *m,r* nomenclature; however, the concept of two distinct types of isotactic dyads within a polymer chain is lost. Secondly, the *m,r* scheme must be used carefully to avoid excessive simplification or even a misinterpretation of the polymer structure. For example,

m m m m m m m m m m m m r m m m m m m m m m m m m

and

m m m m m m m m m m m m r r m m m m m m m m m m m

appear to describe similar polymer structures that differ only by the length of *racemic* additions. As it turns out, different structures, which depend upon the number of *racemic* additions, are obtained with the 0,1 nomenclature:

0 0 0 0 0 0 0 0 0 0 0 0 0 0 0 1 1 1 1 1 1 1 1 1 1 1 1 1 1
m m m m m m m m m m m m m m r m m m m m m m m m m m m m m

and

0 0 0 0 0 0 0 0 0 0 0 0 0 0 0 1 0 0 0 0 0 0 0 0 0 0 0 0 0 0
m m m m m m m m m m m m m m r r m m m m m m m m m m m m m m

The latter is a polymer that contains only a single irregularity among a series of like-handed *meso* additions. The former could be described as a stereoblock polymer of left- and right-handed *meso* additions. The structures of these two polymers are not so similar as possibly inferred from a casual glance at the *m,r* structural formulas.

Both systems of nomenclature have their advantages and will be used jointly throughout this book with the 0,1 system used for its conceptual advantage to replace the stick and ball designations introduced by Bovey in developing the *m,r* system. All of the chemical shift assignments and comonomer distributions will be described with the *m,r* system. Development of the concepts associated with number-average sequence lengths in copolymers and like configurations in homopolymers will necessarily invoke the 0,1 system of nomenclature.

The distribution of *m* and *r* configurations fall under the general heading of polymer *tacticity*. We shall be concerned with three types of quantitative tacticity measurements

(a) distributions of configurations of successive units in vinyl homopolymers such as dyads, triads, tetrads, etc.;
(b) the number-average sequence length of like configurations; and
(c) the number-average sequence lengths of *meso* and *racemic* dyads.

These measurements are necessarily related; consequently, they will be discussed in detail.

1.2 CARBON-13 CHEMICAL SHIFT BEHAVIOR IN VINYL HOMOPOLYMERS

Tacticity measurements require a technique that allows *meso* configurations to be distinguished from *racemic* configurations. With the development of ^1H NMR applications to polymers in solution during the last decades, there were many studies describing vinyl polymer tacticity because the ^1H sensitivity to configurations generally involved dyads, triads, or tetrads. Precise determinations of tacticity, however, were often difficult because the observed spectra represented a composite of complex patterns originating independently from different polymer configurational sequences. The least equivocal NMR tacticity studies involved either high frequency NMR measurements (10), selective deuterations (11, 12) or a combination of these techniques (13).[†]

With the innovation of ^{13}C NMR, polymer tacticity and sequence distribution measurements improved. The sensitivity to configuration increased to five, six, and seven units, complex spectral patterns could be simplified through ^1H heteronuclear decoupling, and chemical shift differences were often large enough that spectral overlap was not a severe problem. In this and the remaining sections of Chapter 1, the reader will be given a background in polymer ^{13}C NMR chemical shift behavior and techniques for spectral assignments. The development of number-average sequence length measurements begins in Chapter 2.

The useful ^{13}C NMR spectral range encountered in most solution polymer spectra is approximately 25 times that found in corresponding ^1H NMR spectra. Information concerning subtle structural features not detected or readily apparent by any other spectroscopic technique are often obvious in ^{13}C NMR spectra. Signals or resonances, obtained from carbon atoms in structurally unique environments, have intensities that are directly proportional to the number of contributing nuclei. Thus ^{13}C NMR spectra contain information concerning the relative numbers and types of polymer nuclei.

Very complex polymer ^{13}C spectral patterns are often observed because of an intrinsic ^{13}C chemical shift sensitivity to structural detail. The ^{13}C spectrum may be complicated further if ^{13}C–^1H spin–spin interactions are allowed which lead to a splitting of lines that could be associated with a single carbon type in a specific environment. A normal ^{13}C NMR spectrum of a polypropylene is shown in Fig. 1.2.[‡] The spectrum contains at least nine lines and, without any other information, it would be difficult to establish unequivocally whether the number of lines shown in Fig. 1.2 arises from ^{13}C–^1H spin–spin coupling, from a multiplicity of ^{13}C environments, or

[†] For an additional background in polymer ^1H applications, the reader is referred to a comprehensive review by Bovey (4).

[‡] Solution polymer NMR spectra are usually obtained at temperatures of 100–150°C where referencing to an internal tetramethylsilane (TMS) standard is impractical. Consequently, hexamethyldisiloxane (HMDS) was used as the internal standard and the chemical shift scale corrected to TMS by adding 2.03 ppm.

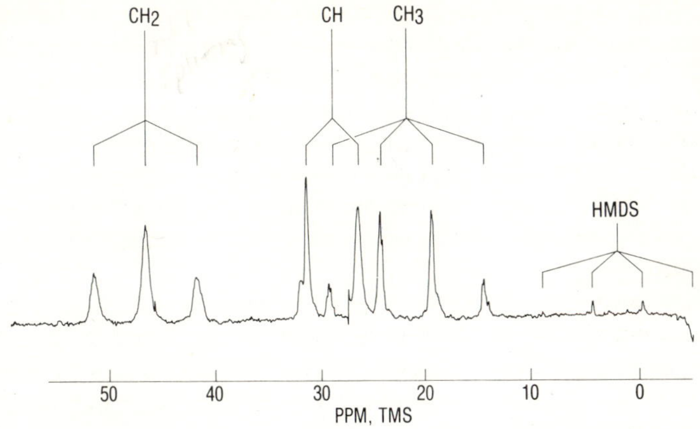

Fig. 1.2. ^{13}C NMR spectrum at 25.2 MHz without proton decoupling of a crystalline polypropylene in 1,2,4-trichlorobenzene at 120°C.

from a combination of both. Fortunately this problem can be resolved through ^{1}H noise heteronuclear decoupling, a double resonance experiment that removes all of the ^{1}H spin coupling with ^{13}C nuclei. An assembly of singlet lines is now left where each can be identified as a chemical shift of a specific polymer carbon nucleus. The number of lines in an ^{1}H decoupled ^{13}C NMR spectrum, therefore, is important and related to the complexity of the structural environments.

An ^{1}H noise decoupled spectrum of the polymer in Fig. 1.2 is shown in Fig. 1.3. Only three lines remain; one for each carbon type found in polypropylene. This result is consistent with that for a polypropylene containing predominantly isotactic sequences where all methyl, methylene, and methine carbons occupy structurally identical environments. In contrast, an amorphous polypropylene with a variety of successive configurational placements has, correspondingly, more structural environments for each carbon type. The ^{13}C NMR spectrum of an amorphous polypropylene is shown in Fig. 1.4. The spectrum is ^{1}H noise decoupled, and contains 17–20 lines that reflect the number of environments for the various types of carbon atoms. A ^{13}C sensitivity to detail is evident, and considerable structural information is available about this amorphous polypropylene if chemical shift assignments can be established for each line in the spectrum.

There are several techniques for assignments; one of the simplest approaches involves the identification of lines from isotactic sequences in an atactic polymer by comparison with the corresponding spectrum of a polymer containing predominantly isotactic sequences. The lines in Fig. 1.4 which correspond to those in Fig. 1.3 are identified by an I. However, even

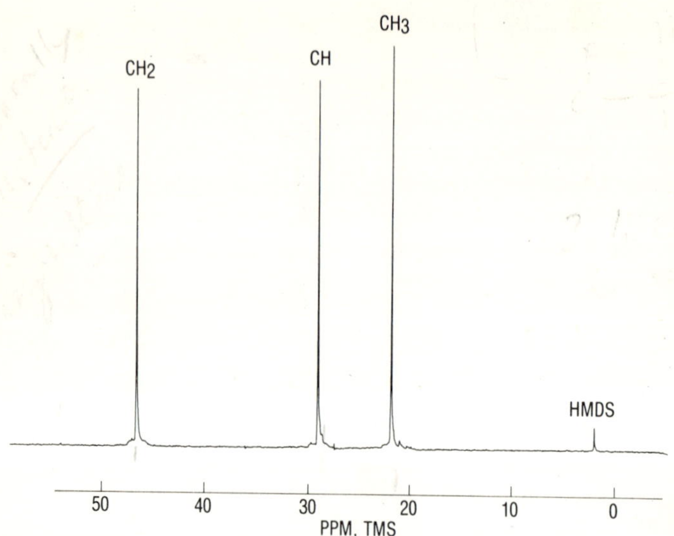

Fig. 1.3. Proton noise decoupled ^{13}C NMR spectrum at 25.2 MHz of a crystalline polypropylene in 1,2,4-trichlorobenzene at 120°C.

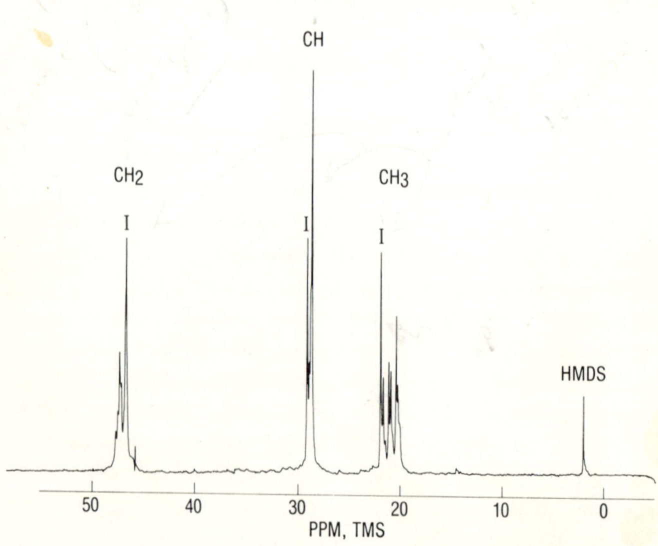

Fig. 1.4. Proton noise decoupled ^{13}C NMR spectrum at 25.2 MHz of an amorphous polypropylene in 1,2,4-trichlorobenzene at 120°C.

with this information, the carbon type has not been specified, that is, whether a resonance arises from either a methyl, methine, or methylene carbon. Secondly, it must be established whether the observed "splittings" as shown in Fig. 1.4 occur because of differences in configurational environments or irregularities in head-to-tail placements.

The chemical shift differences in ^{13}C NMR polymer spectra are related to the number, type, and relative configuration of nearest-neighbor carbon atoms. The three carbon atoms nearest a carbon of interest strongly affect the chemical shift while weaker contributions arise from carbons in the fourth and fifth positions away (14). Sizeable chemical shift differences (10–40 ppm), therefore, can be produced by the skeletal arrangement. Configurational arrangements also lead to chemical shift differences but only by an order of 1 to 2 ppm or less (15). In the following sections, methods are presented that allow identification of a specific resonance source. In particular, it is shown how resonances can be identified for monomer units in the same skeletal arrangement but possessing different configurations.

(a) Off-Resonance Decoupling

Off-resonance decoupling is a technique for identifying the carbon type, that is, whether it is methyl, methylene, methine, or quaternary. In the removal of ^{1}H–^{13}C spin–spin splitting by noise decoupling, information about the carbon type, which is available from line multiplicity, is lost. Proton–carbon spin–spin splitting leads to a quartet for methyl groups, a triplet for methylene carbons, and a doublet for methine carbons. Because both ^{13}C and ^{1}H nuclei have a spin quantum number of $\frac{1}{2}$, and the resonance frequencies are widely separated, the $n + 1$ rule (16) is obeyed for predicting first-order splitting patterns, that is, the line multiplicity is given simply by by the number of interacting, usually directly bonded protons plus one. As seen in Fig. 1.2, ^{13}C NMR spectra taken without ^{1}H noise decoupling can be complex and often difficult to unravel because the ^{1}H–^{13}C coupling constants can be greater than the local chemical shift differences. Although the spectrum in Fig. 1.2 is not exceedingly complex, it did arise from a system of only three major spin types; thus the interpretation of such spectra becomes increasingly ambiguous as the number of lines increases.

The technique of off-resonance decoupling was introduced to overcome the problem of complexity yet allow carbon-type identifications. A single ^{1}H decoupling frequency, which is offset 100–200 Hz from the resonance of interest, is employed as opposed to broad band ^{1}H noise decoupling. The net result is that the ^{1}H–^{13}C splittings are observed but not at the former separations. Multiplet spacings of only a few hertz can be produced that allow an identification of the carbon type. In Fig. 1.5, an off-resonance decoupled spectrum is shown of polypropylene that contains predominantly isotactic sequences. (See Figs. 1.2 and 1.3 for normal and ^{1}H noise decoupled spectra, respectively.) As shown in Fig. 1.5, the methyl resonance is identified through the appearance of a quartet, the methylene resonance by a triplet, and the methine resonance by a doublet. Carbon-type assignments for predominantly isotactic polypropylene are therefore unambiguous.

Off-resonance decoupling is a valuable tool for carbon-type assignments in ^{13}C NMR spectra of most polymers and copolymers. It will not lead, however, to a discrimination of the same carbon type in different structural environments. An interpretation of the results after off-resonance decoupling can still be ambiguous if the lines overlap as shown by the off-resonance decoupled spectrum of the amorphous polypropylene in Fig. 1.6. Other techniques or methods are still needed. Perhaps the most valuable has been assignments based on established chemical shift behavior.

(b) Grant and Paul Empirical Rules

Grant and Paul (14) developed an empirical method for alkane ^{13}C chemical shift assignments based on carbon skeleton arrangements. The method

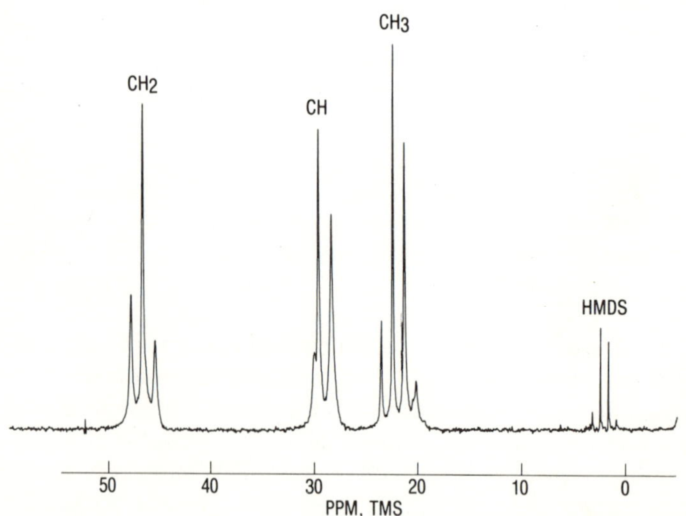

Fig. 1.5. Proton off-resonance decoupled ^{13}C NMR spectrum at 25.2 MHz of a crystalline polypropylene in 1,2,4-trichlorobenzene at 120°C.

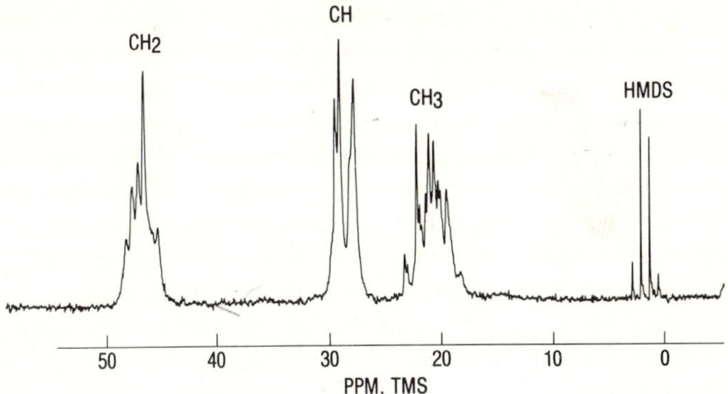

Fig. 1.6. Proton off-resonance decoupled ^{13}C NMR spectrum at 25.2 MHz of an amorphous polypropylene in 1,2,4-trichlorobenzene at 120°C.

is of great value to the practicing NMR spectroscopist because it can be used to discern resonances from carbons of the same type but in different structural environments. In an early study of linear and branched alkane ^{13}C chemical shift behavior, Grant and Paul formulated a set of empirical rules for predicting alkane carbon chemical shifts upon observing that a ^{13}C chemical shift could be dissected into contributions from its immediate carbon neighbors in terms of both bond distance and molecular geometry. Chemical shift contributions to a carbon of interest were detected as far as five bonds away and described by α for the first bonded carbon, β for the second carbon two bonds away, γ for the third, δ for the fourth, and ε for the fifth. The use of these parameters can best be understood through a relatively easy application. For *n*-dodecane, the neighboring contributions to the chemical shift of carbon number six are

$$\underset{1}{\overset{\varepsilon}{CH_3}}-\underset{2}{\overset{\delta}{CH_2}}-\underset{3}{\overset{\gamma}{CH_2}}-\underset{4}{\overset{\beta}{CH_2}}-\underset{5}{\overset{\alpha}{CH_2}}-\underset{6}{\overset{*}{CH_2}}-\overset{\alpha}{CH_2}-\overset{\beta}{CH_2}-\overset{\gamma}{CH_2}-\overset{\delta}{CH_2}-\overset{\varepsilon}{CH_2}-CH_3$$

where

$$\text{chemical shift of } 6 = 2\alpha + 2\beta + 2\gamma + 2\delta + 2\varepsilon + \text{constant} \quad (1.1)$$

Values for α through ε were obtained following a linear regression analysis of previously assigned chemical shifts, set up as above, from a series of linear hydrocarbons. The constant obtained is the linear regression constant and was observed to correspond closely to the chemical shift of methane. This result is expected because the model is based on each carbon representing an alkyl-substituted methane.

No additional terms or parameters other than those described were needed to define the chemical shifts for normal alkanes; branched alkanes, however, could not be so easily described and required additional, so-called corrective terms to account for chemical shifts of carbons associated with branching. The introduction of third and fourth neighboring alkyl groups for any particular carbon atom affect dramatically the steric interactions that cause chemical shift changes. Branched carbons and carbons next to branches show upfield chemical shifts from values calculated with parameters α through ε. To account for this behavior, Grant and Paul introduced *corrective terms* to describe the geometry or type of adjacently bonded carbon atoms. Quaternary, tertiary, secondary, and primary carbon atoms were designated by 4°, 3°, 2°, and 1°, respectively. In the corrective terms, the carbon of interest was listed first and the adjacently bonded carbon was placed in parenthesis. For example, the 3°(2°) corrective term gives the contribution to a methine carbon chemical shift from steric effects associated with an adjacent methylene group. Values were found for the various possible corrective terms that produced an effect upon an observed chemical shift. Some corrective terms such as 3°(1°) were unnecessary because no additional geometrical contributions other than that from α were observed for the chemical shift of a methine carbon directly bonded to a methyl group.

The Grant and Paul parameter values have proven valuable in ^{13}C NMR polymer spectra for distinguishing carbons of the same type but located in different structural environments. However, difficulties not encountered in corresponding alkane studies have been found in obtaining accurate methine carbon chemical shifts. Subsequently, it was shown that better agreement between calculated and observed polymer chemical shifts could be obtained if the corrective terms were modified for polymers (17). Not only were the values for polymer corrective terms different from those of the homologous lower molecular weight alkanes but a temperature dependence was noted. The values of α through ε, however, remained essentially the same for both polymers and alkanes. Parameter values measured from a series of ethylene–1-olefin copolymers and hydrogenated polybutadienes are given in Table 1-1. The polymer corrective terms listed in Table 1-1 are usually sufficient for distinguishing structural features found in most vinyl homopolymers

TABLE 1-1 Grant and Paul Parameter Values[a]

α	8.61 ± 0.18 ppm	3°(2°)	−2.65 ± 0.08 ppm	
β	9.78 ± 0.16	2°(3°)	−2.45 ± 0.17	
γ	−2.88 ± 0.10	1°(3°)	−1.40 ± 0.38	
δ	0.37 ± 0.14	Regression constant[b]		−1.87 ppm
ε	0.06 ± 0.13	Number of observations		56

[a] Measured from polymers in 1,2,4-trichlorobenzene solutions at 125°C (17).
[b] With respect to an internal tetramethylsilane (TMS) standard.

and copolymers. Other parameters, which may be required for highly branched polymers and not listed in the table, may be obtained from the original study of Grant and Paul or from a later related study of Carman *et al.* (18). It should be remembered, however, that values for these corrective terms were not obtained from polymers but from alkanes at temperatures below 50°C.

As an example of the value of the Grant and Paul empirical rules for assignments according to the structural environment, the methyl, methylene, and methine chemical shifts for polypropylene are calculated. (Chemical shift contributions are designated specifically for the methine carbon.)

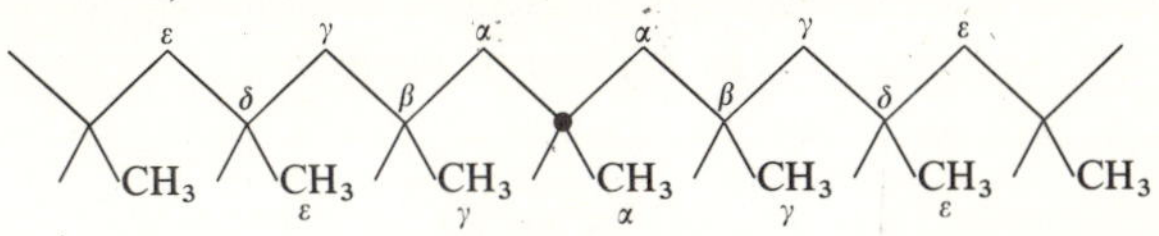

$$\mathrm{CH} = 3\alpha + 2\beta + 4\gamma + 2\delta + 4\varepsilon + 2(3^\circ(2^\circ)) + (-1.87) = 27.68\ \mathrm{ppm} \quad (1.2)$$

$$\mathrm{CH_2} = 2\alpha + 4\beta + 2\gamma + 4\delta + 2\varepsilon + 2(2^\circ(3^\circ)) + (-1.87) = 45.41\ \mathrm{ppm} \quad (1.3)$$

$$\mathrm{CH_3} = \alpha + 2\beta + 2\gamma + 4\delta + 2\varepsilon + 1^\circ(3^\circ) + (-1.87) = 20.74\ \mathrm{ppm} \quad (1.4)$$

The polypropylene chemical shifts, calculated with the Grant and Paul parameters, closely match those obtained experimentally. Amorphous polypropylene gives methyl chemical shifts in a range of 20 to 22 ppm, methylene chemical shifts from 44 to 47 ppm, and methine chemical shifts from 26 to 29 ppm. Correspondingly, predominantly isotactic polypropylene has only a single methyl resonance at 21.80 ppm, a single methylene resonance at 46.52 ppm, and likewise, only a single methine resonance at 28.50 ppm. In both samples, there are chemical shift contributions from steric effects associated with configurational arrangements that were not included in the calculation. It is evident from the spectrum of amorphous polypropylene in Fig. 1.4 that configurational contributions varied because additional splittings not exceeding 1 to 2 ppm occurred in the vicinity of each chemical shift predicted for head-to-tail monomer additions. Configurational contributions to carbon chemical shifts have not yet been included among the Grant and Paul parameter terms; however, splittings related to configurational arrangements can be identified quite easily because it appears superimposed upon much larger chemical shift differences caused by carbon skeleton arrangements (15).

Up to this point, only chemical shifts associated with propylene unit head-to-tail arrangements have been considered. Skeletal arrangements such as head-to-head and tail-to-tail monomer placements (inversion) should give chemical shifts that are quite different from head-to-tail arrangements because the parameter values for α and β carbon contributions are approximately 10 ppm. In fact, substantial chemical shift differences are predicted for methyl and methine carbons in isolated head-to-head sequences and for methylene carbons in isolated tail-to-tail sequences as demonstrated with calculated chemical shifts using the Grant and Paul parameters found in Table 1-1.

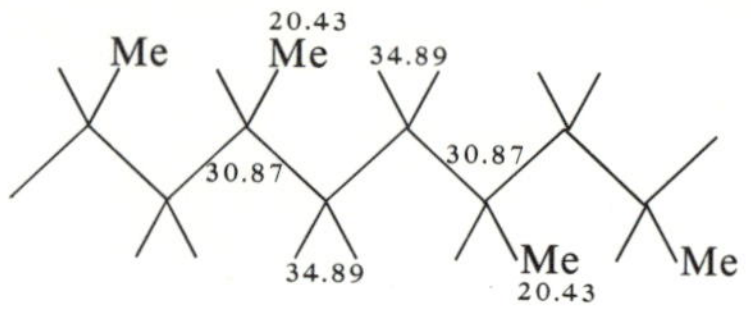

Me 20.43 Me 34.89 30.87 30.87 34.89 Me 20.43 Me

Tail-to-tail monomer placements

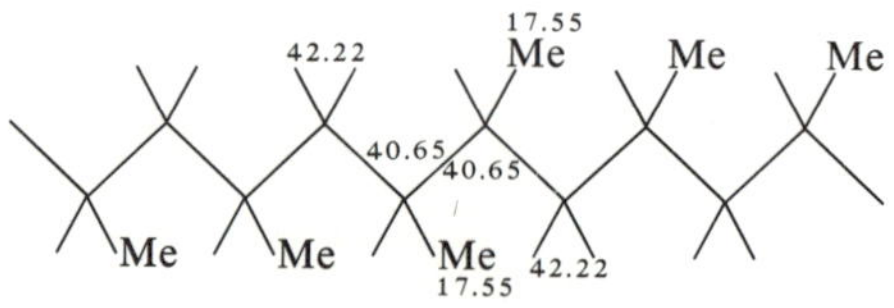

42.22 17.55 Me Me Me 40.65 40.65 Me Me Me 17.55 42.22

Head-to-head monomer placements

A spectrum of a polypropylene containing resonances from head-to-head and tail-to-tail monomer sequences is shown in Fig. 1.7. The resonances not seen in previous polypropylene spectra (see Figs. 1.3 and 1.4) can be assigned from these calculated chemical shifts because close agreement was obtained. Detailed assignments are given in Fig. 1.7. Additional lines are also observed among these resonances because configurational contributions are once again imposed upon skeleton contributions to each chemical shift.

Although polypropylene has been discussed in detail, the Grant and Paul technique can be extended to any vinyl polymer. In some cases, contributions to the polymer backbone chemical shifts from substituents other than alkyl groups must be determined with appropriate model compounds. Parameter values have been obtained for phenyl (19) and acetate groups (20) using the Grant and Paul approach. Other substituent parameters are given by Levy (2) and can be applied to various vinyl polymers. Attention has not been given to parameters that define steric relationships in vinyl polymers; such studies will likely occur when assignments are substantiated. An extension of the Grant and Paul parameters to account for configurational differences among methyl groups in polypropylene (22) has apparently proven successful for assignments; however it is not general. Unfortunately, the number of

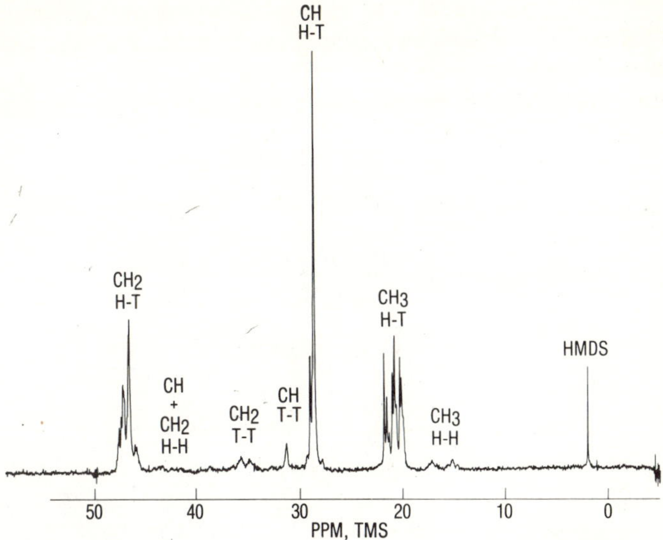

Fig. 1.7. Proton noise decoupled ^{13}C NMR spectrum at 25.2 MHz of a polypropylene containing inverted monomer additions in 1,2,4-trichlorobenzene at 120°C.

additional parameters required to describe configurational relationships often exceeds the degrees of freedom available for the analysis, thus a more rigorous approach may be through appropriate model compounds and polymers.

1.3 CONFIGURATIONAL SEQUENCE LENGTHS AND CARBON-13 LINE MULTIPLICITY

Before discussing configurational assignments in vinyl homopolymer ^{13}C spectra any further, let us examine NMR chemical shift sensitivity with respect to the total number of configurational arrangements possible for a particular sequence length. The numbers of lines observed in spectra of amorphous polypropylenes for carbon chemical shifts from head-to-tail arrangements occur because of an intrinsic chemical shift sensitivity toward certain configurational sequence lengths. This chemical shift sensitivity also varies according to the position of a carbon atom within a monomer unit because of symmetry. Thus different multiplets will be observed for each carbon atom in units of the same configuration. In polypropylene, for example, the methylene carbons are directly bonded to two asymmetric centers:

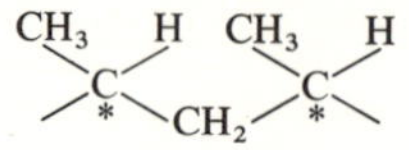

Thus the simplest form of configurational sensitivity involves dyads, that is,

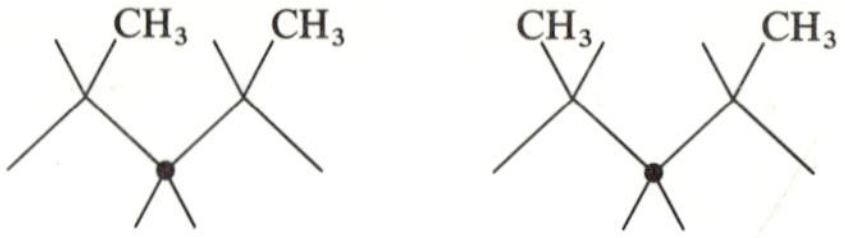

meso *racemic*

and proceeds to tetrad, hexad, etc., as additional near neighbors affect the chemical shift sensitivity. For methine or methyl carbons in polypropylene, the simplest configurational sensitivity involves at least triads because the asymmetric carbons are now next nearest neighbors instead of nearest neighbors.

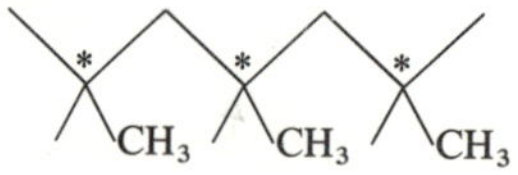

The configurational sensitivity, therefore, begins with triad and proceeds to pentad, heptad, etc., as additional near neighbors affect the local screening environments.

For an interpretation of polymer ^{13}C spectra, we need a procedure for identifying the chemical shift sensitivity to configuration length for each type of carbon resonance. We can begin by counting the number of unique resonances possible for each sequence length. For example, there are two dyads, three triads, six tetrads, and ten pentads. The configuration length detected, therefore is often determined by simply noting the number of lines in a ^{13}C NMR spectrum for a specific carbon type. In Fig. 1.4, extensive splittings are evident for each carbon type in the spectrum of an amorphous polypropylene. There are too many lines for either a dyad or triad sensitivity. Each line must be identified to take full advantage of the structural information present because the relative intensities reflect correspondingly the structural distribution of the average polymer chain.

Let us now examine the method for determining the number of unique configurations possible for each sequence length in more detail. A simple dyad sensitivity produces only two lines, one for *meso* and another for *racemic* configurations although there are four dyads if individual unit configurations are considered. In this case,

$$00 = 11 = m \qquad \text{and} \qquad 01 = 10 = r$$

For those carbons exhibiting a triad sensitivity, there are 2^3 or eight possible arrangements if, once again, we consider individual unit configurations. These are

$$111 = 000 = mm, \qquad 110 = 001 = mr$$

$$011 = 100 = rm, \qquad 010 = 101 = rr$$

However, four triads are not observed because the *mr* triad cannot be distinguished from the *rm* triad. After a 180° rotation the two are the same; that is, neither chain direction nor chirality can be specified. The number of unique triads, therefore, reduces to only three, *mm*, *mr*, and *rr*. For ideally random polymers, where the number of *m* dyads is the same as the number of *r* dyads, the triad distribution will be 1:2:1 for the ratio of *mm* to *mr* to *rr*. If such a triad sensitivity were shown for the amorphous polypropylene in Fig. 1.4, the methyl and methine resonances would consist of three peaks with 1:2:1 relative intensities.

The next higher level of NMR configurational sensitivity is tetrad sequences for methylene carbons and pentad sequences for either methine or side-chain carbons. In a manner similar to the arguments above, it can be shown that there are six unique tetrads and ten unique pentads. In practice, the number of unique arrangements for triads, pentads, etc., can be more easily predicted with the *m,r* nomenclature as shown in Fig. 1.8 where

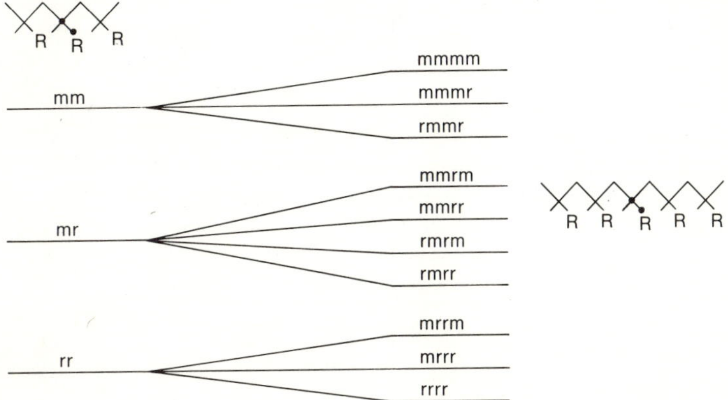

Fig. 1.8. Progression from triad to pentad sequences for methine and side-chain carbons in vinyl polymers.

the ten pentads are divided into three groups: those with *mm* centers (three), those with *mr* centers (four), and those with *rr* centers (three). An analogous situation exists for methylene carbons that begins with dyads and proceeds to tetrads, hexads, etc., as the chemical shift sensitivity increases to include additional near neighbors. Figure 1.9 contains the progression in chemical shift splitting for methylene carbons from dyads to hexads that leads to the number of unique lines possible for a particular chemical shift sensitivity.

An expanded view of the amorphous polypropylene spectrum in Fig. 1.4 is shown in Fig. 1.10. The easiest spectral region to interpret is the methyl region. A pentad chemical shift sensitivity is clearly indicated with the *mm*, *mr*, and *rr* regions readily identified. The number of lines observed in the *rr* region suggests a heptad sensitivity for some of the sequences with *rr* centers. This result may not be surprising because the chemical shift sensitivity is a reflection of the near-neighbor interactions possible for each particular con-

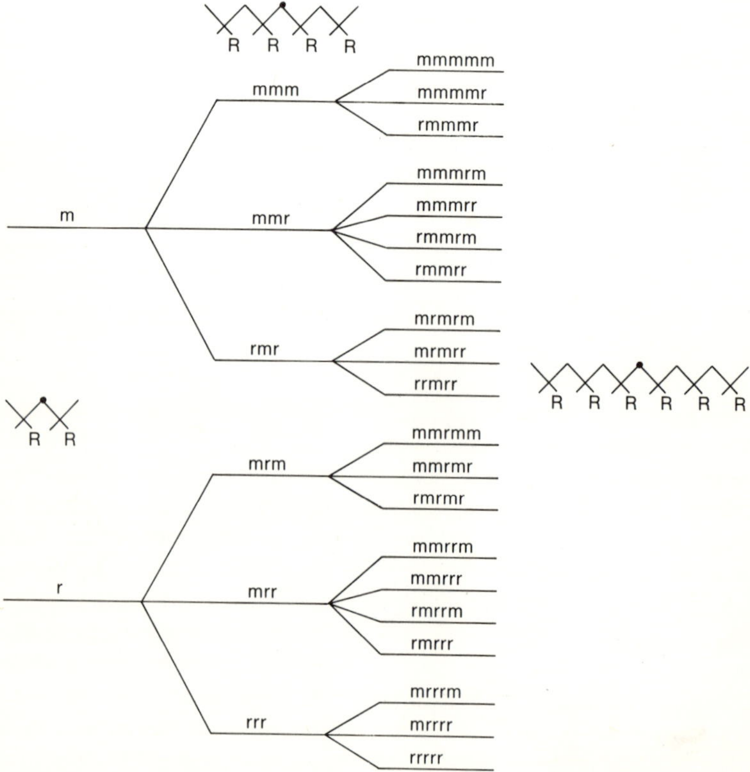

Fig. 1.9. Progression from dyad to tetrad to hexad sequences for methylene carbons in vinyl polymers.

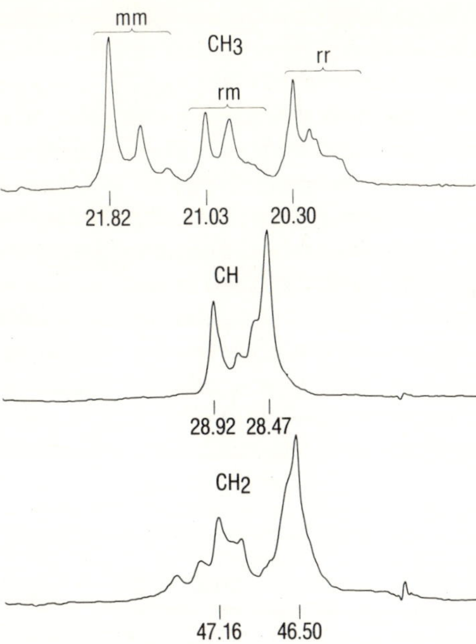

Fig. 1.10. Expanded methyl, methine, and methylene regions of the amorphous polypropylene shown in Fig. 1.4.

formation. A specific triad or pentad may not possess the required properties to produce a chemical shift difference or, on the other hand, conformational averaging may not give rise to a chemical shift difference. Apparently only the *rr*-centered heptads are producing observable chemical shift differences for this amorphous polypropylene.

From symmetry considerations, the methine carbon resonances will also be sensitive to the odd sequence lengths. Only four lines could be detected in the methine region shown in Fig. 1.10. The basic chemical shift sensitivity, therefore, cannot be attributed to just triads. The first two methine lines probably represent pentads because the relative intensities are the same as those observed for the first two lines of the methyl region. The third and fourth methine lines are likely combinations of the remaining eight pentads. The methine chemical shift sensitivity, therefore, is pentad although a complete set of ten pentads is not resolved. Very little configurational information can be obtained from the methine ^{13}C resonances; however this has been offset by the excellent resolution displayed in the methyl region.

The methylene resonances are sensitive to even-numbered configurational sequences. In Fig. 1.10, a polypropylene methylene region is shown which contains more than six but less than 20 lines. The chemical shift sensitivity can best be described as hexad with observable differences not shown by some of the hexad sequences.

In principle, chemical shift sensitivities of any ^{13}C NMR polymer spectrum can be established by inspection of the number of lines. Supporting evidence should be obtained because the number of observed lines may be deceptively simple as was the case for the methine region of the amorphous polypropylene. Of the vinyl polymers studied so far, the methylene regions are generally characterized by tetrad and hexad chemical shift sensitivities, and the methine regions by triad and pentad sensitivities.

In addition to defining the progression in number of lines, Figs. 1.8 and 1.9 also give the relationships among relative concentrations of configurational sequences. For example, if the relative concentrations of the ten methyl pentads are known for the amorphous polypropylene, then the relative triad concentrations can be determined by summing the pentad concentrations that have the same triad center, that is,

$$mmmm + mmmr + rmmr = mm \tag{1.5}$$

$$mmrm + mmrr + rmrm + rmrr = mr \tag{1.6}$$

$$mrrm + mrrr + rrrr = rr \tag{1.7}$$

Analogous relationships exist for heptads to pentads, etc., although the heptad lines are not shown in Fig. 1.8. Corresponding relationships for hexads to tetrads to dyads are shown in Fig. 1.9.

Relationships between configurational sequences of different lengths have been defined previously by Bovey (7, 11, p. 19). These are listed in Table 1-2 and are referred to hereafter as the "necessary *n*-ad" relationships. Other

TABLE 1-2 Necessary Relationships between Relative Concentrations of Configurational Sequences of Different Lengths[a]

Triad–dyad	$mm + \frac{1}{2}mr$	$= m$
	$rr + \frac{1}{2}mr$	$= r$
Tetrad–triad	$mmm + \frac{1}{2}mmr$	$= mm$
	$rmr + \frac{1}{2}mmr + mrm + \frac{1}{2}mrr$	$= mr$
	$rrr + 1/r\ mrr$	$= rr$
Pentad–tetrad	$mmmm + \frac{1}{2}mmmr$	$= mmm$
	$rmmr + \frac{1}{2}mmmr + \frac{1}{2}mmrm + \frac{1}{2}mmrr$	$= mmr$
	$\frac{1}{2}mrmr + \frac{1}{2}rmrr$	$= rmr$
	$\frac{1}{2}mrmr + \frac{1}{2}mmrm$	$= mrm$
	$mrrm + \frac{1}{2}mrrr + \frac{1}{2}mmrr + \frac{1}{2}rmrr$	$= mrr$
	$rrrr + \frac{1}{2}mrrr$	$= rrr$

[a] See Bovey (4, p. 19).

useful relationships exist between configurational sequences of the same length (19). These are

$$rmmr + \tfrac{1}{2}mmmr = \tfrac{1}{2}mmrm + \tfrac{1}{2}mmrr \quad (1.8)$$

$$mrrm + \tfrac{1}{2}mrrr = \tfrac{1}{2}rmrr + \tfrac{1}{2}mmrr \quad (1.9)$$

for pentads and

$$rmr + \tfrac{1}{2}mmr = mrm + \tfrac{1}{2}mrr \quad (1.10)$$

for tetrads. These various relationships among relative concentrations of configurational sequences are available for testing assignments within a given spectral region and for testing the consistency of assignments between methylene and methine carbons where different chemical shift sensitivities are shown. Finally, in quantitative studies where detailed assignments are not possible, peak intensities can be combined to produce a concentration for a shorter sequence length. The ultimate reduction, of course, would be to relative concentrations of *meso* and *racemic* dyads.

1.4 CONFIGURATIONAL ASSIGNMENTS IN POLYPROPYLENE

Chemical shift multiplicities, observed in ^{13}C NMR spectra of vinyl polymer methylene and methine carbons in identical head-to-tail arrangements are caused by intrinsic differences among configurational relationships for successive monomer units. Therefore, these differences in ^{13}C NMR chemical shifts are a manifestation of the dissimilarities in the conformational averaging process for a given series of monomer unit configurations as compared to others that occur along the polymer chain. A prediction of chemical shift behavior based on conformational properties, however, is complicated by the possibility that differences in conformational properties do not necessarily lead to chemical shift differences, as discussed in Section 1.3. The magnitude of a given chemical shift is also governed by magnetic properties of neighboring nuclei that can be optimized or minimized through certain angular relationships, which may be independent of the geometries that produce either favorable or unfavorable conformations. Steric crowding does produce upfield chemical shifts and, to an extent, chemical shifts can be correlated with conformational properties; however, the ^{13}C chemical shift behavior that appears to be general for one vinyl polymer cannot necessarily be extended to a second type of vinyl system.

For these reasons model compounds, although valuable, should be used judiciously.

Probably the best approach to assignments involves a comparison of spectra from polymers of known configurational structure. As in previous sections, this discussion will be limited to polypropylene assignments; however, the techniques used can be applied to other vinyl polymers.

Predominantly isotactic and syndiotactic reference polypropylenes can be prepared with Zeigler–Natta catalyst system for assignment of various *mm—m* and *rr—r* configurations (23). These polymers, however, lead to only two of ten possible methyl assignments, two of four methine assignments, and only two of at least six methylene assignments. Stehling and Knox (24) have indicated that *epimerization* (25) of predominantly isotactic and syndiotactic polypropylenes can be used for five more methyl assignments. Briefly, an addition of 1% dicumylperoxide plus 4% tris(2,3-dibromopropyl)phosphate to a stereochemically pure polypropylene causes random, well-spaced inversions of monomer unit configurations if conversions are deliberately kept low. Thus epimerization produces the following changes in the structure of a predominantly isotactic polymer:

0 0 0 0 0 0 0 0 0 0 0 0 0 0 0 0 0 0 0 0 0 0 0 0 0 0 0 0 0

↓ (1) tertiary hydrogen extraction
(2) chain transfer leading to inversion

0 0 0 0 0 0 0 0 0 0 0 0 0 0 1 0 0 0 0 0 0 0 0 0 0 0 0 0 0

and introduces the pentad sequences

00001	(2)	*mmmr*
00010	(2)	*mmrr*
00100	(1)	*mrrm*

Correspondingly, epimerization of a predominantly syndiotactic polymer,

0 1 0 1 0 1 0 1 0 1 0 1 0 1 0 1 0 1 0 1 0 1 0 1 0 1 0 1 0 1

↓ (1) tertiary hydrogen extraction
(2) chain transfer leading to inversion

0 1 0 1 0 1 0 1 0 1 0 1 0 0 0 1 0 1 0 1 0 1 0 1 0 1 0 1 0 1

produces the unique pentad sequences

01000	(2)	*rrmm*
10100	(2)	*rrrm*
10001	(1)	*rmmr*

Upon completion of the epimerization experiments, three new resonances in a 2:2:1 ratio will be present in the methyl region of the ^{13}C NMR spectra of both the predominantly isotactic and predominantly syndiotactic polypropylenes. In each case, *rmmr* and *mrrm* can be assigned uniquely because these resonances will show one-half the intensity of the other two. The *mmrr* resonance can also be uniquely assigned because it is the only commonly produced pentad from the two experiments. Finally, *mrrr* and *rmmm* are assigned by default. These results lead to unequivocal assignments for seven of the ten methyl resonances in amorphous polypropylenes.

Before continuing with a discussion of the final three ^{13}C methyl assignments, let us examine the new tetrad sequences associated with methylene carbons from epimerization of both isotactic and syndiotactic polymers:

$$\text{isotactic} \rightarrow \begin{matrix} 0001 \quad (2) \\ (mmr) \\ \\ 0010 \quad (2) \\ (mrr) \end{matrix} \qquad \text{syndiotactic} \rightarrow \begin{matrix} 0100 \quad (2) \\ (rrm) \\ \\ 1000 \quad (2) \\ (rmm) \end{matrix}$$

As shown above, the same new tetrads are produced from both isotactic and syndiotactic polymers. (The reaction scheme used to predict the pentads formed upon epimerization can also be used to demonstrate why the same tetrads are produced for the methylene carbons.) A second disadvantage in this particular application is that unique assignments are not obtained because the resonance intensities will be the same. Epimerization can only support methylene tetrad assignments made by some other technique.

The final three polypropylene methyl assignments, *mmrm*, *rmrr*, and *rmrm*, are possible from a consideration of the necessary pentad–pentad relations (Eqs. (1.8) and (1.9)),

$$2rmmr + mmmr = mmrm + mmrr \tag{1.11}$$

$$2mrrm + mrrr = mmrr + rmrr \tag{1.12}$$

utilizing the relative intensities of each of the methyl resonances (see Table 5-1). In each equation only one unknown pentad appears; therefore, the relative concentration of the pentad can be obtained and compared with the observed relative intensities. An assignment would be based on the existence of discriminating differences among the observed intensities for *mmrm*, *rmrr*, and *rmrm*. On a basis of the pentad–pentad relationships, the *mr*-centered pentads in Fig. 1.10 are assigned in order of *mmrr*, *rmrr*, *mmrm*, and *rmrm* from low to high field (24). These assignments are tentative because of uncertainties in the *mrrm* and *mrrr* concentrations as determined from the ^{13}C NMR spectrum.

Model compounds also provide a useful means for assignments and Zambelli *et al.* (26) synthesized labeled ^{13}C model compounds, 3(s),5(r),7(rs),-9(rs),11(rs),13(r),15(s)-heptamethylheptadecane (compound A) and a mixture of A with 3(s),5(s),7(rs),9(rs),11(rs),13(r),15(s)-heptamethylheptadecane for this purpose. The model compounds led to nine pentad assignments in order from low to high field: *mmmm*, *mmmr*, *rmmr*, *mmrr*, *mmrm* + *rmrr*, *rmrm*, *rrrr*, *rrrm*, and *mrrm*, which are in agreement with the seven previous assignments by epimerization and lead to a positive assignment for *rmrm*. Assignments for *mmrm* and *rmrr* cannot be obtained unequivocally from this model compound because these resonances overlap in the model compound while they do not appear to overlap in ^{13}C polymer spectra. Indications are present however, from low temperature spectra of this model compound that *mmrm* occurs at a lower field than *rmrr* (26a). Until the question of the number of pentad resonances identified in the *mr* region is resolved, the final two assignments *mmrm* and *rmrr* should remain tentative. This *mmrm* and *rmrr* assignment, shown in Fig. 1.11, also agrees with an observed additivity relationship that takes into consideration possible steric interactions (22). These assignments, however, are not in agreement with the best fit according to the necessary pentad–pentad relationships given by Eqs. (1.11) and (1.12). Until spectra, perhaps at higher magnetic fields, clearly

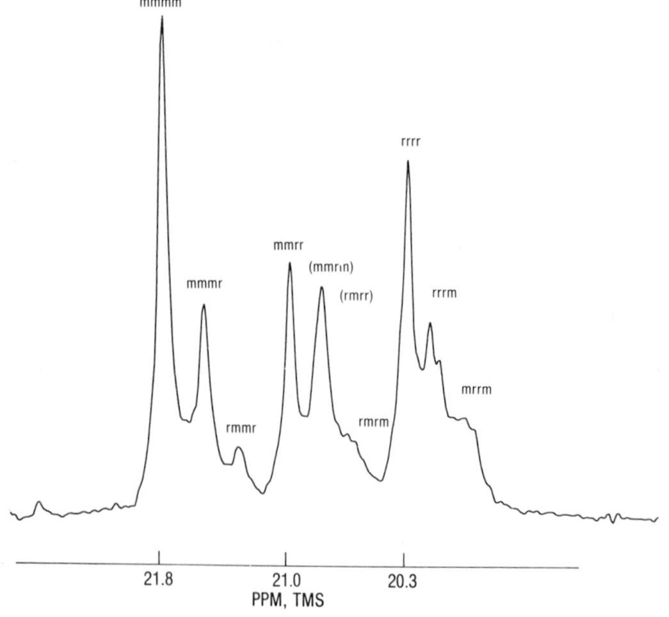

Fig. 1.11. Expanded methyl region of the amorphous polypropylene shown in Fig. 1.4.

delineate the four *mr*-centered pentads, these final two assignments should remain tentative. As we shall see later in an analysis of the number-average sequence lengths of like configurations in polypropylene, an identification of the methyl resonances according to triad centers is sufficient for a complete analysis.

In summary, problems associated with ^{13}C NMR assignments are not easily resolved although techniques are available for readily discerning the carbon-type and skeletal arrangements. Specific, configurational assignments have been made from established additive trends among the chemical shift data but are more reliably made if based on model compounds and polymers. Model compound approaches, however, must be used judiciously because the conformational properties of the polymer chain are probably not reproduced.

An additional assignment technique, just recently applied to polymer spectra, involves what is called "suppression of resonance".[†] Overlap among resonances can be suppressed or eliminated entirely if the overlapping resonances have sufficiently different spin–lattice relaxation times. Through a $180°$–τ–$90°$ pulse sequence, overlap can be removed through an appropriate choice in delay time (τ) between the $180°$ and $90°$ pulses. Carbons with different spin–lattice relaxation times give resonances which null at different points in time after a $180°$ pulse. Dramatic results have been achieved by Gerken and Ritchey[‡] in ^{13}C polymer spectra where overlap occurred between resonances from carbons of different types.

In Chapter 4, Markovian statistical approaches to configurational assignments will be discussed. We shall now assume that established assignments are available and shall be concerned with the quantitative use of ^{13}C NMR data. With information concerning the sequencing of polymer chain units, we possess the unique advantage of determining more than the simple monomer distribution. Number-average sequence lengths for like monomer additions can also be measured whether the system is a vinyl homopolymer consisting of *meso* and *racemic* dyads or a copolymer with different structural units.

[†] The author is indebted to Professor William M. Ritchey for pointing out the technique of suppression of resonance during a review of the manuscript.

[‡] T. A. Gerken and W. M. Ritchey, private communication.

2 *Number-Average Sequence Lengths in Vinyl Polymers*

In Chapter 1, the ^{13}C NMR chemical shift behavior in vinyl polymers was interpreted from a standpoint of resonance multiplicities and underlying causes. Chemical shift differences from skeletal arrangements—head-to-tail, head-to-head, and tail-to-tail monomer additions—were distinguished from those arising from configurational differences among like skeletal arrangements. Considerable information, which is dependent upon assignments and sorting of spectral data into its most useful form, is available from ^{13}C polymer spectra because the relative peak intensities for the majority of polymers do reflect in a corresponding way the relative abundance of structural entities.

This chapter deals with concepts associated with a conversion of a monomer distribution to a number-average sequence length. As discussed in Chapter 1, a vinyl homopolymer from a configurational viewpoint can be considered as a copolymer of successive units of either the same or opposite handedness. For example,

1 0 0 0 1 0 1 1 0 1 0 0 1 1 1 1 0 1 1 1 0 0 0 0 1 0

where series of *like* configurations occur as runs of 0 additions (0, 00, 000, etc.) or as runs of 1 additions (1, 11, 111, etc.). Although chirality cannot be identified, the number-average run length of *like* configurations is an important quantity for polymer characterization (27). The above chain can also be defined as a copolymer of *racemic* and *meso* dyads,

r m m r r r m r r r m r m m m r r m m r m m m r r

where two monomer units are now needed to define configuration and any inference of chirality is avoided. Number-average sequence lengths, defined for both *meso* and *racemic* configurations, differ from the corresponding run lengths of *like* configurations because two units are required to define relative additions. A number-average sequence length can therefore be measured in two ways: the number-average sequence length of *like* configurations, and the number-average sequence length of *meso* and *racemic* additions. The former definition is consistent with that for homopolymers because only one type of run, interrupted by opposite placements, is involved while the latter definition is totally consistent with a copolymer framework. In neither case can run lengths of opposite handedness be determined because absolute configurations cannot be assigned. As stated previously, the concept of chirality is defined relatively within a given chain. The chirality of successive units in adjacent chains can correspond to either configuration by symmetry operations of end-to-end rotation and appropriate translation. In the following sections, the sequence lengths of stereochemical additions in vinyl homopolymers are characterized by both approaches, the number-average sequence length of *like* configurations and the number-average sequence lengths of *meso* and *racemic* additions.

2.1 NUMBER-AVERAGE SEQUENCE LENGTHS OF LIKE CONFIGURATIONS

Let us examine the number-average sequence length of runs of like configurations designated by either 0 or 1 and terminated on either end with an opposite configuration. In the chain segment examined previously, the sequences are defined as

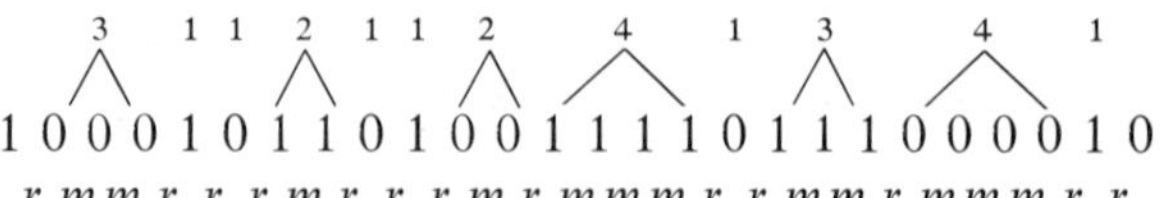

where there are six runs of one unit each, two runs containing two like units, two runs containing three like units, and two runs containing four like units. The number-average sequence length of like successive additions for this particular chain segment is

$$\frac{6(1) + 2(2) + 2(3) + 2(4)}{6 + 2 + 2 + 2} = 2.0 \tag{2.1}$$

(The *m,r* designations are also included above for the convenience of the reader.)

The number-average sequence length $\bar{n}$ for like configurations can be described mathematically in a general form as

$$\bar{n} = \frac{\sum_{i=0}^{i=n} iN_{1(0)_i1} + \sum_{i=0}^{i=n} iN_{0(1)_i0}}{\sum_{i=1}^{i=n} N_{1(0)_i1} + \sum_{i=1}^{i=n} N_{0(1)_i0}} \tag{2.2}$$

for an average chain containing N runs of successive like configurations from 1 to n. It is necessary to include sums over runs of both 0 and 1 units in Eq. (2.2); however, because the 0's and 1's can be interchanged, a shorthand version can be written using only one of the sums. For brevity, a 0-centered nomenclature will be adopted where, by definition,

$$N_{101} = N_{101} + N_{010} = N_{rr} \tag{2.3}$$

$$N_{1001} = N_{1001} + N_{0110} = N_{rmr} \tag{2.4}$$

$$\vdots$$

$$N_{1(0)_n1} = N_{1(0)_n1} + N_{0(1)_n0} = N_{r(m)_{n-1}r} \tag{2.5}$$

Equation (2.2) can, therefore, be written more simply as

$$\bar{n} = \sum_{i=0}^{i=n} iN_{1(0)_i1} \bigg/ \sum_{i=1}^{i=n} N_{1(0)_i1} \tag{2.6}$$

A number-average sequence length for like configurations can be determined from Eq. (2.6) provided information concerning the concentrations or number per average chain of each successive placement is known. Nuclear magnetic resonance measurements provide such data because there is a structural sensitivity to at least dyads of monomer units. Just how to put this information to its best use to determine number-average sequence lengths is not always immediately apparent; consequently, a variety of approaches appear in the literature. One of the purposes of this book is to provide a general approach to the problem of sequence determinations applicable to both homopolymers and copolymers and to present a method that is easy to put into practice.

Although a complete set of concentrations describing each sequence length appears necessary for a number-average sequence length determination, one needs to know only the relative concentrations of successive monomer units in terms of sequence lengths of at least two units. As will be shown shortly, the number-average sequence length of like configurations is given by the reciprocal of the *racemic* dyad concentration that can be obtained from any monomer distribution, two units and longer. Therefore, derivations relating monomer distributions to number-average sequence lengths can begin with dyad, triad, tetrad, or any higher distribution. The triad distribution is chosen initially because the derivation can be made without employing any of the necessary n-ad relationships.

A triad distribution, (000), (001), and (101), is defined as

$$(000) = \frac{N_{000}}{N_{000} + N_{001} + N_{101}} \tag{2.7}$$

$$(001) = \frac{N_{001}}{N_{000} + N_{001} + N_{101}} \tag{2.8}$$

$$(101) = \frac{N_{101}}{N_{000} + N_{001} + N_{101}} \tag{2.9}$$

where the total number of triads per average chain, N_{000}, is counted from like sequences three units in length and longer by

$$N_{000} = \sum_{j=0}^{j=n} jN_{10(0)_j01} \tag{2.10}$$

Note that j counts the number of triads per run and n is the longest sequence of triads in the polymer chain. The N_{001} triad counts the number of runs of like units, two and longer,

$$N_{001} = 2\sum_{j=0}^{j=n} N_{10(0)_j01} \tag{2.11}$$

Finally, the N_{101} triad represents a unique combination of 0 and 1 units and is simply

$$N_{101} = N_{101} \tag{2.12}$$

Equation (2.6), which defines the number-average sequence length for like configurations, can be expressed in a form that allows the number-average sequence length to be defined in terms of the triad distribution, that is,

$$\bar{n} = \frac{N_{101} + \sum_{j=0}^{j=n} (j+2)N_{10(0)_j01}}{N_{101} + \sum_{j=0}^{j=n} N_{10(0)_j01}} \tag{2.13}$$

which leads to

$$\bar{n} = \frac{N_{000} + N_{100} + N_{101}}{N_{101} + \frac{1}{2}N_{001}} \tag{2.14}$$

upon substitution of Eqs. (2.10)–(2.12). If defined with respect to relative concentrations, (000), (001), etc., Eq. (2.14) becomes

$$\bar{n} = ((101) + \tfrac{1}{2}(001))^{-1} \tag{2.15}$$

or

$$\bar{n} = ((rr) + \tfrac{1}{2}(mr))^{-1} \tag{2.16}$$

With the dyad–triad necessary relationship in Table 1-2, Eq. (2.16) reduces to

$$\bar{n} = 1/(r) \tag{2.17}$$

This number-average sequence length can, in an analogous manner, be determined from any comonomer distribution. The required equation can be developed from Eq. (2.17) with the appropriate n-ad relationship or derived independently in the manner above.

For derivations involving higher-order monomer distributions, the necessary n-ad relationships must be used in the expanded form of Eq. (2.6) to obtain an expression relating the number-average sequence length to a uniform monomer distribution. Although there is an easier derivation, it may be worthwhile to examine how a number-average sequence length is obtained from a pentad distribution using Eq. (2.6) because triad–tetrad–pentad necessary relationships must be invoked initially in the derivation. For a pentad distribution, Eq. (2.6) expands to

$$\bar{n} = \frac{N_{101} + 2N_{1001} + 3N_{10001} + \sum_{j=0}^{j=n} (j + 4)N_{100(0)_j001}}{N_{101} + N_{1001} + N_{10001} + \sum_{j=0}^{j=n} N_{100(0)_j001}} \tag{2.18}$$

Note that j now defines the number of pentads per run and n the longest pentad sequence. The 101 triad and 1001 tetrad concentrations can be defined in terms of pentad concentrations through triad–pentad and tetrad–pentad necessary relationships:

$$N_{101} = N_{01010} + N_{01011} + N_{11011} \tag{2.19}$$

and

$$N_{1001} = \tfrac{1}{2}N_{10010} + \tfrac{1}{2}N_{10011} \tag{2.20}$$

In Eq. (2.19), the 101 triad concentration is obtained from a sum over all the pentads that have a 101 center. The relationship given by Eq. (2.20) (see Table 1-2) is not so easy to see; the factor $\frac{1}{2}$ arises because the 1001 sequence is counted twice by the pentads in each of the longer sequences involving 1001 placements:

0 1 0 0 1 0   0 1 0 0 1 1   1 1 0 0 1 1

The remaining terms in Eq. (2.18) can be defined by the same approach used in the triad derivation, that is,

$$\sum_{j=0}^{j=n} jN_{100(0)_j001} = N_{00000} \tag{2.21}$$

$$2\sum_{j=0}^{j=n} N_{100(0)_j001} = N_{00001} \tag{2.22}$$

A substitution of Eqs. (2.19)–(2.22) into Eq. (2.18) gives the number-average sequence length of like configurations as a function of pentad concentrations only.

$$\bar{n} = \frac{N_{01010} + N_{01011} + N_{11011} + N_{10010} + N_{10011} + 3N_{10001} + 2N_{00001} + N_{00000}}{N_{01010} + N_{01011} + N_{11011} + \frac{1}{2}N_{10010} + \frac{1}{2}N_{10011} + N_{10001} + \frac{1}{2}N_{00001}} \tag{2.23}$$

or in the m,r notation:

$$\bar{n} = \frac{(rrrr) + (rrrm) + (mrrm) + (rmrr) + (rmrm) + 3(rmmr) + 2(mmmr) + (mmmm)}{(rrrr) + (rrrm) + (mrrm) + \frac{1}{2}(rmrr) + \frac{1}{2}(rmrm) + (rmmr) + \frac{1}{2}(mmmr)} \tag{2.24}$$

As it turns out, only eight of the ten possible pentad concentrations (see Fig. 1.9) appear in the numerator of Eq. (2.24). If desired, this result can be rectified through the use of the necessary pentad–pentad relationship,

$$2(rmmr) + (mmmr) = (mmrm) + (mmrr) \tag{2.25}$$

to give

$$\bar{n} = \left[(rrrr) + (rrrm) + (mrrm) + \tfrac{1}{2}(rmrr) + \tfrac{1}{2}(rmrm) + (rmmr) + \tfrac{1}{2}(mmmr)\right]^{-1} \tag{2.26}$$

Finally, Eq. (2.26) can be reduced to the corresponding equation for triads,

Eq. (2.16), through introduction of the triad–pentad necessary relationships from Table 1-2,

$$(rr) = (rrrr) + (rrrm) + (mrrm) \tag{2.27}$$

$$\tfrac{1}{2}(mr) = \tfrac{1}{2}(rmrr) + \tfrac{1}{2}(rmrm) + (rmmr) + \tfrac{1}{2}(mmmr) \tag{2.28}$$

Although the number-average sequence length of like configurations is given simply by $1/(r)$, expressions defined in terms of higher-order monomer combinations offer an easy approach for equating a monomer distribution to a number-average sequence length. The only other equation that may be useful in polymer configurational analyses would be that from a tetrad distribution,

$$\bar{n} = \left[(rrr) + \tfrac{1}{2}(rrm) + (rmr) + \tfrac{1}{2}(mmr)\right]^{-1} \tag{2.29}$$

which can be derived independently from either Eq. (2.17) or Eq. (2.6) in a manner analogous to the derivations of Eqs. (2.16) and (2.26).

This characterization of vinyl polymers through sequence lengths of like configurations leads to a description of the polymer structure in terms of runs of like additions only. An ideally random polymer, where the mole fractions $(m) = (r) = \frac{1}{2}$, has a number-average sequence length of 2.0. For syndiotactic polymers, which have alternating configurations, a number-average sequence length of 1.0 is obtained for like configurations. For any polymer with a number-average sequence length higher than two, there are more *meso* than *racemic* configurations.

2.2 NUMBER-AVERAGE SEQUENCE LENGTHS OF *MESO* AND *RACEMIC* ADDITIONS

As we discussed earlier, number-average sequence lengths can also be measured in terms of *meso* and *racemic* additions. For the same polymer chain segment used in the development of number-average sequence lengths of like configurations,

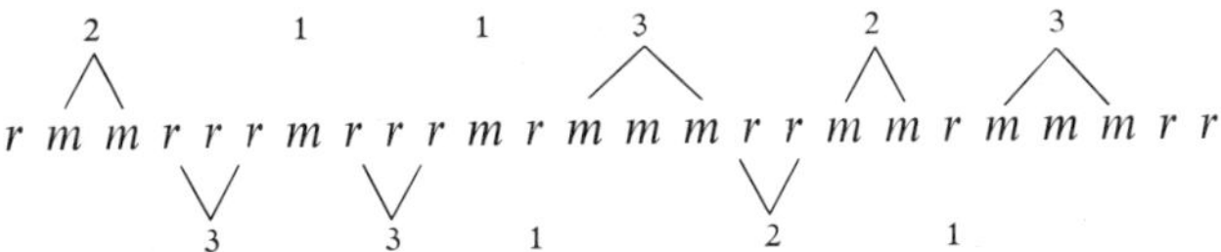

the number-average sequence length of *m* configurations is

$$\frac{2(1) + 2(2) + 2(3)}{2 + 2 + 2} = 2.0 \tag{2.30}$$

and of *r* configurations,

$$\frac{2(1) + 1(2) + 2(3)}{2 + 1 + 2} = 2.0 \tag{2.31}$$

The derivation for number-average sequence lengths from a particular monomer distribution follows closely that for like configurations. The number-average sequence for runs of *meso* additions, $\bar{n}_{\mathrm{m}}$, is

$$\bar{n}_{\mathrm{m}} = \frac{\sum_{i=0}^{i=n} iN_{r(m)_i r}}{\sum_{i=1}^{i=n} N_{r(m)_i r}} \tag{2.32}$$

and for *racemic* configurations, $\bar{n}_{\mathrm{r}}$,

$$\bar{n}_{\mathrm{r}} = \frac{\sum_{i=0}^{i=n} iN_{m(r)_i m}}{\sum_{i=1}^{i=n} N_{m(r)_i m}} \tag{2.33}$$

To determine the number-average sequence length of either *meso* or *racemic* additions, we must detect the *mr* sequence independently from the *mm* and *rr* placements. Therefore, a triad sensitivity is required to define such a polymer chain and the numbers of *mm*, *mr*, and *rr* placements per average chain are given by

$$N_{\mathrm{mm}} = \sum_{j=0}^{j=n} jN_{rm(m)_j r} \tag{2.34}$$

$$N_{\mathrm{rr}} = \sum_{j=0}^{j=n} jN_{mr(r)_j m} \tag{2.35}$$

$$N_{\mathrm{mr}} = \sum_{j=0}^{j=n} N_{rm(m)_j r} + \sum_{j=0}^{j=n} N_{mr(r)_j m} \tag{2.36}$$

Equations (2.32) and (2.33) can be expanded to a form similar to Eqs. (2.34)–(2.36):

$$\bar{n}_{\mathrm{m}} = \frac{\sum_{j=0}^{j=n} (j + 1)N_{rm(m)_j r}}{\sum_{j=0}^{j=n} N_{rm(m)_j r}} \tag{2.37}$$

and

$$\bar{n}_{\mathrm{r}} = \frac{\sum_{j=0}^{j=n} (j + 1)N_{mr(r)_j m}}{\sum_{j=0}^{j=n} N_{mr(r)_j m}} \tag{2.38}$$

Substitution of Eqs. (2.34)–(2.36) into Eqs. (2.37) and (2.38) with the realization that for high *m,r* alterations,

$$N_{\mathrm{mr}} = 2 \sum_{j=0}^{j=n} N_{rm(m)_j r} = 2 \sum_{j=0}^{j=n} N_{mr(r)_j m} \tag{2.39}$$

leads to

$$\bar{n}_{\mathrm{m}} = \frac{(mm) + \frac{1}{2}(mr)}{\frac{1}{2}(mr)} \tag{2.40}$$

and

$$\bar{n}_{\mathrm{r}} = \frac{(rr) + \frac{1}{2}(mr)}{\frac{1}{2}(mr)} \tag{2.41}$$

Corresponding equations can be derived for tetrads, pentads, etc. The final result for a tetrad monomer distribution is

$$\bar{n}_{\mathrm{m}} = \frac{(rmr) + (mmr) + (mmm)}{(rmr) + \frac{1}{2}(mmr)} \tag{2.42}$$

and

$$\bar{n}_{\mathrm{r}} = \frac{(mrm) + (rrm) + (rrr)}{(mrm) + \frac{1}{2}(mrr)} \tag{2.43}$$

As an exercise, the reader may wish to derive Eqs. (2.42) and (2.43) either from Eqs. (2.40) and (2.41) or from Eqs. (2.32) and (2.33).

The characterization of tacticity through the *meso*, *racemic* copolymer model requires a higher chemical shift sensitivity from NMR data than does the model for average sequence lengths for like configurations because a dyad distribution can be used for the latter case but not for the former. This is not necessarily a disadvantage since ^{13}C NMR data typically show tetrad, pentad, and hexad chemical shift sensitivities. For polymers with a stereoblock structure, it would be advantageous to use the *meso*, *racemic* copolymer model. Finally, as was the case for number-average sequence lengths of like configurations, ideally random polymers give

$$\bar{n}_{\mathrm{m}} = \bar{n}_{\mathrm{r}} = 2.0 \tag{2.44}$$

2.3 NMR MEASUREMENTS OF NUMBER-AVERAGE SEQUENCE LENGTHS

An NMR spectrum represents the net accumulations of millions of nuclear signals from various polymer chains that are discriminated according to their structural environments. If recorded under appropriate experimental conditions (see Chapter 5), the relative intensities of these nuclear signals will reflect correspondingly the polymer structural distribution. Because an accumulation of signals from all of the various polymer molecules occurs, the resultant NMR spectrum represents an "average" polymer molecule. Any quantitative interpretation of NMR spectral data therefore necessarily applies to an average structure and does not reflect possible differences in the structural distribution that may occur either within or between polymer chains.

The purpose of this section is to discuss briefly quantitative structural descriptions of NMR resonance areas that can be used to calculate either monomer distributions or number-average sequence lengths. A discussion of the experimental aspects of such NMR measurements will be given in Chapter 5.

The NMR technique is ideally suited for measurements of either monomer distributions or number-average sequence lengths. First, solution spectra can be recorded in such a way that the resonance areas are directly proportional to the number of contributing groups. There are no extinction coefficients as in infrared or ultraviolet analyses but a standard proportionality constant k that includes all of the variables in an NMR analysis that become fixed when a specific set of experimental conditions is selected. A description of NMR resonance areas, therefore, parallels the development of the equations for number-average sequence lengths because resonance areas also count the number of contributing groups. For example, with a triad sensitivity, the resonance areas of *mm*, *mr*, and *rr* sequences are described by

$$I_{\mathrm{mm}} = kN_{\mathrm{mm}} = k \sum_{j=0}^{j=n} jN_{10(0)_j01} \tag{2.45}$$

$$I_{\mathrm{mr}} = kN_{\mathrm{mr}} = 2k \sum_{j=0}^{j=n} N_{10(0)_j01} \tag{2.46}$$

$$I_{\mathrm{rr}} = kN_{\mathrm{rr}} = kN_{101} \tag{2.47}$$

where k includes such factors as the number of molecules in the vicinity of the receiver coil, instrumental response, and recorder attenuation. It is unnecessary to evaluate k because every resonance has the same value for k.

Although the procedure for calculating number-average sequence lengths from any uniform monomer distribution was shown in the preceding section, one can describe each resonance intensity according to its exhibited chemical shift sensitivity, as in Eqs. (2.13) or (2.18) (or a suitable expanded form). An example is given in Chapter 3 for hydrogenated polybutadienes.

Thus it is unnecessary to obtain a complete monomer distribution before calculating a number-average sequence length.

A second advantage of NMR in sequence distribution measurements relates specifically to ^{13}C NMR Fourier transform experiments. Data can be collected through free induction decay (FID) accumulations until a desired signal-to-noise ratio is achieved. Because ^{13}C NMR spectra are usually noise decoupled, each peak in the final spectrum represents a specific polymer entity analogous to a gas chromatogram of a mixture of chemical compounds. The techniques for area measurements applied in gas–liquid chromatography analyses are strictly applicable in ^{13}C NMR spectra. Curve resolving using Lorentzian shapes is a particularly sound approach to NMR area measurements.

A possible difficulty in ^{13}C NMR quantitative measurements may be found in the nuclear Overhauser effect (28). As a consequence of ^{1}H noise decoupling, energy transfer can occur between ^{1}H and ^{13}C spin levels and lead to an enhancement of the observed ^{13}C signals. In low molecular weight organic molecules, which have relatively nonrestricted molecular motions, the nuclear Overhauser effect (NOE) varies according to carbon type or location (29). Schaefer (30) has shown that corresponding differences in NOE's do not occur for ^{13}C nuclei in solution spectra of polymer molecules. Restricted molecular motions have led to NOE's that are generally constant throughout the polymer molecule.

Finally, it may be beneficial to determine the number-average sequence lengths in the two polypropylenes (see Figs. 1.3 and 1.4) examined earlier in Chapter 1. Relative areas, obtained through curve resolving using Lorentzian shapes, are given in Table 2-1 (27). The data in Table 2-1 describe triad and pentad monomer distributions and can be used to determine both the number-average sequence length of like configurations and number-average sequence lengths of *meso* and *racemic* configurations. For polypropylene, the triad assignments are confirmed; however, pentad assignments *mmrm* and *rmrr* are tentative, as discussed in Chapter 1. Number-average sequence length determinations, therefore, are made with more confidence from a triad distribution. (Results from the triad distribution may be compared with those from the pentad distribution as a check for internal consistency of the pentad assignments.) Number-average sequence lengths, using the triad distributions in Table 2-1 and Eqs. (2.16), (2.40), and (2.41), for the amorphous polypropylene, are

$$\bar{n} = 2.0, \qquad \bar{n}_{\mathrm{m}} = 3.3, \qquad \bar{n}_{\mathrm{r}} = 3.4$$

Similarly, the crystalline polypropylene gave

$$\bar{n} = 33, \qquad \bar{n}_{\mathrm{m}} = 49, \qquad \bar{n}_{\mathrm{r}} = 1.5$$

TABLE 2-1 Relative Areas of Ten ^{13}C Methyl Resonances[a]

Triad	Pentad[b]	Amorphous polypropylene		Crystalline polypropylene	
mm	*mmmm*	0.35	0.19	0.95	0.85
	mmmr		0.12		0.08
	rmmr		0.04		0.02
mr	*mmrr*	0.30	0.11	0.04	0.03
	mmrm		0.15		0.01
	rmrr		0.02		0.00
	rmrm		0.02		0.00
rr	*rrrr*	0.36	0.16	0.01	0.00
	rrrm		0.14		0.00
	mrrm		0.06		0.01

[a] From spectra of amorphous and crystalline polypropylenes obtained by curve resolving using Lorentzian peak shapes (27).

[b] See Chapter 1, pp. 23–26, for a discussion of the methyl pentad assignments.

The error in the analysis of a crystalline polypropylene could naturally be quite high because relative differences in peak sizes used in the analysis were approximately 100 to 1. A 50% error in the area measurement of the smaller peaks could result in a 100% difference in the values for number-average sequence lengths. Note that the amorphous polypropylene is not an ideally random polymer; although $\bar{n} = 2.0$, values for $\bar{n}_m$ and $\bar{n}_r$ are not 2.0 as required by Eq. (2.44). If taken alone, the result for number-average sequence lengths of like configurations could be misleading because a value of 2.0 was obtained fortuitously. An example of a nearly ideally random polymer is given in Section 6.8. An average structure for this amorphous polypropylene which satisfies these sequence length requirements is

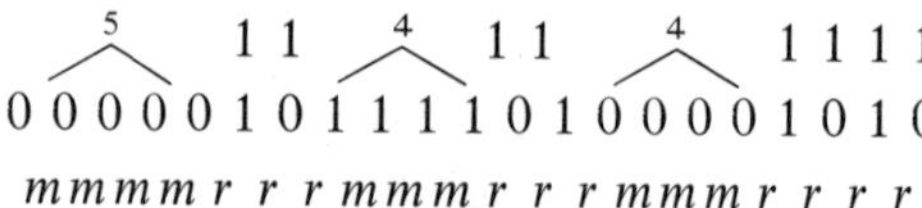

3 Number-Average Sequence Lengths in Copolymers and Terpolymers

In Chapter 2, number-average sequence lengths for like configurations and for *meso* and *racemic* additions in vinyl polymers were derived as a function of specific monomer distributions, that is, consecutive monomer combinations expressed as either dyads, triads, tetrads, pentads, etc. Many of the principles developed for like versus opposite configurations also apply to corresponding analyses in copolymers and terpolymers. (In fact, equations developed for number-average sequence lengths of *meso* and *racemic* additions in vinyl homopolymers are identical to those for copolymers.) A similar nomenclature is used, 0 and 1 denote structurally different monomer units in a copolymer while 0, 1, and 2 identify the different monomer units in a terpolymer. The only difference between the designations for copolymers and those for vinyl polymers is that no degeneracies are encountered such as the equivalency of the 00 and 11 dyads.

The backbone carbon chemical shift sensitivity found in ^{13}C NMR spectra of most copolymers and terpolymers will depend upon the distance in numbers of carbon-atoms between methine carbons from adjacent or nearby units. For copolymers or terpolymers containing butadiene units where additions could be linear or branched, the chemical shift sensitivity is either dyad or triad and occasionally tetrad. Copolymers or terpolymers containing vinyl monomer units can give high chemical shift sensitivities because the methylene and methine carbons can alternate as in vinyl homopolymers. We shall consider those copolymers where an essentially dyad or triad chemical shift sensitivity is observed and use those ^{13}C resonances that can be related to dyad or triad concentrations of monomer units. Even in these cases, the analysis can become very complicated because there are six unique triads in copolymers and 18 in terpolymers. The most desirable descriptions would be in terms of dyads of monomer units because copolymers have only three unique dyad combinations and terpolymers have six.

3.1 NUMBER-AVERAGE SEQUENCE LENGTHS IN COPOLYMERS AND DYAD CONCENTRATIONS

Any copolymer can be described sequentially by a series of 0's and 1's where 0 represents one monomer unit and 1 represents the other. As was the case for vinyl homopolymers, this succession of 0's and 1's can be subdivided into combinations of two, three, four, etc., units. In the previous discussion, the dyad description was said to be most appropriate for copolymers because the ^{13}C chemical shift sensitivity is generally dyad or triad. With information available about connecting unit concentrations, number-average sequence lengths can be determined.

Three types of dyads are possible for a copolymer, the homogeneous dyads 00 and 11, and the heterogeneous dyad 01 or 10. Because each monomer unit begins one dyad and ends another, the homogeneous and heterogeneous dyads are counted from sequence lengths of 1 to n as follows:

$$N_{00} = \sum_{j=0}^{j=n} jN_{10(0)_j1} \tag{3.1}$$

$$N_{11} = \sum_{j=0}^{j=n} jN_{01(1)_j0} \tag{3.2}$$

$$N_{01} + N_{10} = \sum_{j=0}^{j=n} N_{01(1)_j0} + \sum_{j=0}^{j=n} N_{10(0)_j1} + 1 \tag{3.3}$$

There is only one heterogeneous dyad because the chain direction cannot be specified. These dyads are therefore described by either N_{01} or N_{10} which represents the total heterogeneous dyad concentration. In subsequent discussions, N_{01} is used throughout as the total number of heterogeneous dyads per average chain.

The relationship between the total number of 0 and 1 runs and the N_{01} dyad concentration, defined by Eq. (3.3), is not easy to visualize and can best be understood through an examination of the model chain segment

0-centered sequences

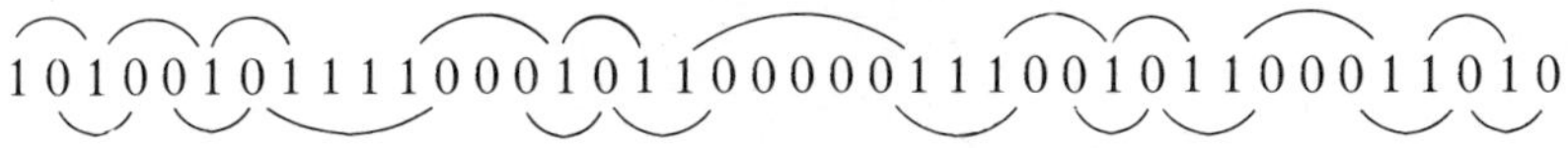

1-centered sequences

The above model has ten 0 runs and ten 1 runs and, therefore, 21 heterogeneous dyads as defined by Eq. (3.3).

For any chain beginning with one unit and ending with the other unit, the number of 1 runs will equal the number of 0 runs, that is,

$$\sum_{j=0}^{j=n} N_{01(1)_j0} = \sum_{j=0}^{j=n} N_{10(0)_j1} \tag{3.4}$$

This circumstance is not true for those chains starting and ending with the same unit. For chains that start and end with a 0 unit,

$$\sum_{j=0}^{j=n} N_{01(1)_j0} = \sum_{j=0}^{j=n} N_{10(0)_j1} + 1 \tag{3.5}$$

and for chains beginning and ending with 1 units,

$$\sum_{j=0}^{j=n} N_{01(1)_j0} = \sum_{j=0}^{j=n} N_{10(0)_j1} - 1 \tag{3.6}$$

Equations (3.4)–(3.6) arise, in part, because of our definition of a run, that is, a succession of one or more like monomer additions terminated on either end by the other monomer addition. The above definitions are not valid for AB or ABA block copolymers where only a few A,B alternations occur. For random copolymers with high numbers of 0,1 alternations, the number of 0-centered runs versus 1-centered runs is defined by the statistical average, Eq. (3.4), and the heterogeneous dyad concentration becomes

$$N_{01} = 2\sum_{j=0}^{j=n} N_{10(0)_j1} = 2\sum_{j=0}^{j=n} N_{01(1)_j0} \tag{3.7}$$

Finally, the number-average sequence lengths for runs of 0 and 1 units are given by

$$\bar{n}_0 = \sum_{i=0}^{i=n} iN_{1(0)_i1} \Big/ \sum_{i=1}^{i=n} N_{1(0)_i1} = \sum_{j=0}^{j=n} (j+1)N_{10(0)_j1} \Big/ \sum_{j=0}^{j=n} N_{10(0)_j1} \quad (3.8)$$

and

$$\bar{n}_1 = \sum_{i=0}^{i=n} iN_{0(1)_i0} \Big/ \sum_{i=1}^{i=n} N_{0(1)_i0} = \sum_{j=0}^{j=n} (j+1)N_{01(1)_j0} \Big/ \sum_{j=0}^{j=n} N_{01(1)_j0} \quad (3.9)$$

A substitution of Eqs. (3.1), (3.2), and (3.7) into these expressions for the number-average sequence lengths leads to the following expressions as a function of dyad concentrations only:

$$\bar{n}_0 = \frac{N_{00} + \frac{1}{2}N_{01}}{\frac{1}{2}N_{01}} \quad (3.10)$$

and

$$\bar{n}_1 = \frac{N_{11} + \frac{1}{2}N_{01}}{\frac{1}{2}N_{01}} \quad (3.11)$$

For the model chain segment used in our discussion of copolymer number-average sequence lengths, $N_{00} = 10$ and $N_{01} = 21$, which leads to a number-average sequence length of 1.95. A result of 2.0 can be obtained with Eq. (3.8) since there are five runs of one unit each, two with two units, two with three units, and one with five units. (Precisely the same results are obtained with Eqs. (3.10), (3.11) and (3.8), (3.9) for chains that start and end with the same unit.) In developing Eqs. (3.10) and (3.11) for copolymer analyses, we have repeated the derivation for the *m,r* copolymer model for vinyl polymers. The final results for a sequence distribution of any length will be in exactly the same form.

Equations (3.10) and (3.11) can also be derived intuitively because the number-average sequence length is simply the number of units of a specific type divided by the number of runs containing that unit. Therefore, in the derivation of Eqs. (3.10) and (3.11), we have also demonstrated that

$$(0) = (00) + \tfrac{1}{2}(01) \quad (3.12)$$

and

$$(1) = (11) + \tfrac{1}{2}(01) \quad (3.13)$$

We are already aware that the number of sequences is $\frac{1}{2}(01)$ from Eq. (3.7). Higher-order equations could be developed from this point by substituting with the necessary relationships between dyads, triads, etc., as was the case for the vinyl homopolymers.

3.2 COPOLYMER NUMBER-AVERAGE SEQUENCE LENGTHS AND TRIAD CONCENTRATIONS

The derivation for number-average sequence lengths defined only in terms of triad concentrations follows basically the same arguments used in the dyad derivation. An intuitive approach can be developed as follows: The total number of 0 units is equal to the triad concentrations $N_{101} + N_{001} + N_{000}$ because these triads represent all of the possible arrangements where 0 is the center unit. The number of runs is simply $N_{101} + \frac{1}{2}N_{001}$. The factor $\frac{1}{2}$ appears because there are two N_{001} triads for each run containing two or more consecutive 0's. The number-average sequence length, therefore, is

$$\bar{n}_0 = \frac{N_{101} + N_{001} + N_{000}}{N_{101} + \frac{1}{2}N_{001}} \tag{3.14}$$

Similarly, for 1 sequences,

$$\bar{n}_1 = \frac{N_{010} + N_{110} + N_{111}}{N_{010} + \frac{1}{2}N_{110}} \tag{3.15}$$

It is instructive to consider also the derivation of Eqs. (3.14) and (3.15) from a standpoint of "counting" triads as a function of sequence length, that is,

$$N_{000} = \sum_{j=0}^{j=n} jN_{10(0)_j01} \tag{3.16}$$

$$N_{001} = 2\sum_{j=0}^{j=n} N_{10(0)_j01} \tag{3.17}$$

$$N_{110} = 2\sum_{j=0}^{j=n} N_{01(1)_j10} \tag{3.18}$$

$$N_{111} = \sum_{j=0}^{j=n} jN_{01(1)_j10} \tag{3.19}$$

which upon substitution into

$$\bar{n}_0 = \frac{N_{101} + \sum_{j=0}^{j=n} (j+2)N_{10(0)_j01}}{N_{101} + \sum_{j=0}^{j=n} N_{10(0)_j01}} \tag{3.20}$$

and

$$\bar{n}_1 = \frac{N_{010} + \sum_{j=0}^{j=n} (j+2)N_{01(1)_j10}}{N_{010} + \sum_{j=0}^{j=n} N_{01(1)_j10}} \tag{3.21}$$

leads to Eqs. (3.14) and (3.15). Note that the derivation for the number-average sequence length in terms of a triad distribution requires no assumptions concerning the number of alternations per chain.

Because the number-average sequence length for either 0 or 1 additions is simply given by

$$\bar{n}_0 = (0)/\tfrac{1}{2}(01) \tag{3.22}$$

and

$$\bar{n}_1 = (1)/\tfrac{1}{2}(01) \tag{3.23}$$

a comparison of Eqs. (3.22), (3.23) with (3.10), (3.11), and (3.14), (3.15) leads to the necessary relationships,

$$N_0 = N_{00} + \tfrac{1}{2}N_{01} = N_{101} + N_{100} + N_{000} \tag{3.24}$$

$$N_1 = N_{11} + \tfrac{1}{2}N_{01} = N_{010} + N_{110} + N_{111} \tag{3.25}$$

$$N_{01} = 2N_{010} + N_{110} = 2N_{101} + N_{001} \tag{3.26}$$

Two other relationships can be obtained as follows:

$$N_{00} = \sum_{i=0}^{i=n} iN_{10(0)_i1} \tag{3.27}$$

$$= \sum_{j=0}^{j=n} (j+1)N_{10(0)_j01} \tag{3.28}$$

$$= \sum_{j=0}^{j=n} jN_{10(0)_j01} + \sum_{j=0}^{j=n} N_{10(0)_j01} \tag{3.29}$$

which from Eqs. (3.16) and (3.17) lead to

$$N_{00} = N_{000} + \tfrac{1}{2}N_{001} \tag{3.30}$$

Similarly,

$$N_{11} = N_{111} + \tfrac{1}{2}N_{110} \tag{3.31}$$

As an exercise, the reader should derive Eq. (3.26) from Eq. (3.7), and any other relationships that may be desired. The necessary relationships are a natural byproduct of derivations for number-average sequence length that represents an advantage of the mathematical as opposed to the intuitive approach. One of the bonuses of the dyad approach for sequence distribution analyses is that the N_{01} dyad can be determined from either the 0- or 1-centered triads. This fact can become important in analyses where either poor resolution or ambiguous assignments has occurred for one set of triads. Before beginning a corresponding derivation for terpolymers, let us examine an analysis of sequence distributions in hydrogenated polybutadienes where these principles are demonstrated.

3.3 NUMBER-AVERAGE SEQUENCE LENGTHS IN HYDROGENATED POLYBUTADIENES

Polybutadienes, although homopolymers from a polymerization viewpoint are structurally copolymers after hydrogenation because the monomer additions can exist in two different configurations. Butadiene monomers insert into growing polymer chains as either *cis*- or *trans*-1,4 additions, or 1,2- or vinyl additions (31). Hydrogenation removes all *cis–trans* differences to produce only two kinds of backbone units, either straight chain (1,4) or ethyl branched (1,2). Carbon-13 NMR spectra of three hydrogenated polybutadienes containing 179 ethyl branches per 1000 carbons (72% 1,2 additions), 125 branches per 1000 carbons (50% 1,2 additions), and 65 branches per 1000 carbons (26% 1,2 additions) are shown in Figs. 3.1–3.3, respectively. In each case, the resonances are numbered from 1 to 19 from low to high

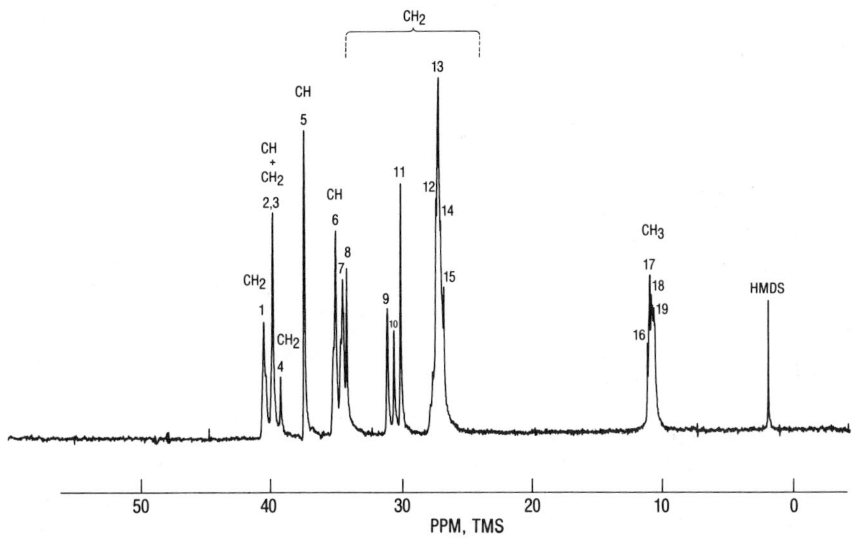

Fig. 3.1. Proton noise decoupled ^{13}C NMR spectrum at 25.2 MHz of a hydrogenated polybutadiene (72% 1,2 additions) in 1,2,4-trichlorobenzene at 120°C.

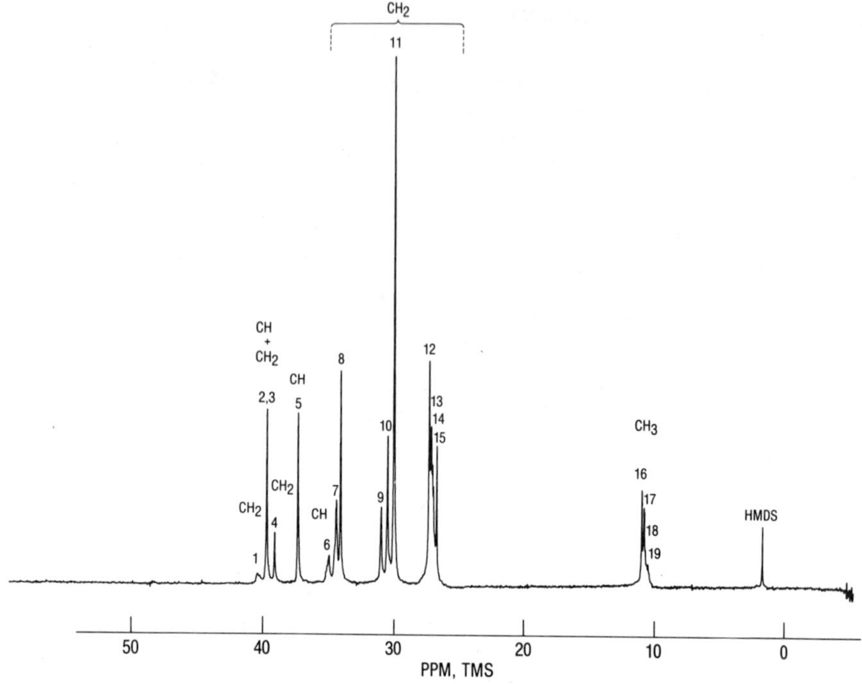

Fig. 3.2. Proton noise decoupled ^{13}C NMR spectrum at 25.2 MHz of a hydrogenated polybutadiene (50% 1,2 additions) in 1,2,4-trichlorobenzene at 120°C.

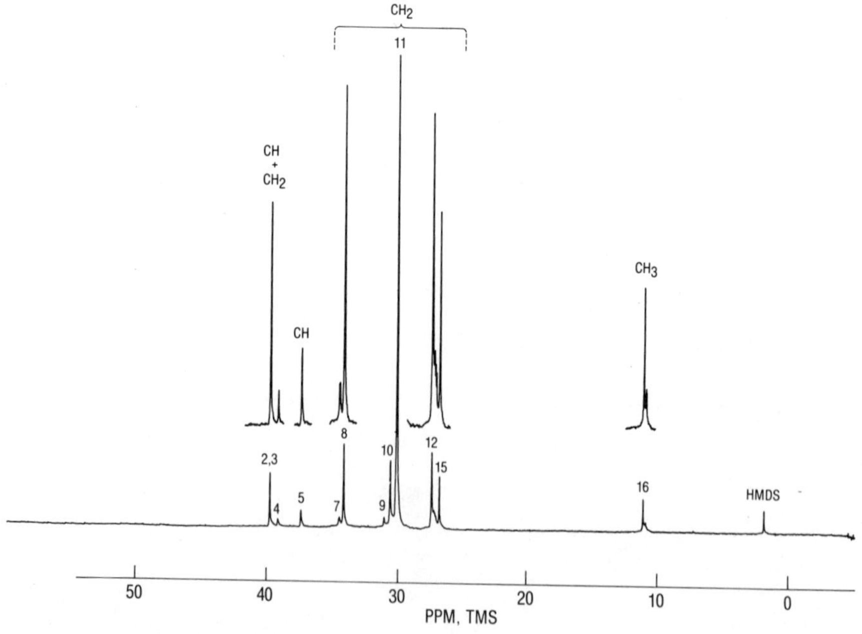

Fig. 3.3. Proton noise decoupled ^{13}C NMR spectrum at 25.2 MHz of a hydrogenated polybutadiene (26% 1,2 additions) in 1,2,4-trichlorobenzene at 120°C.

TABLE 3-1 Equations for Resonance Areas 1–12, 15, and 16[a]

Equation	Assignment[b]
$I_1 = k \sum_{i=2}^{i=n} (i-1) N_{01(1)_i 10}$	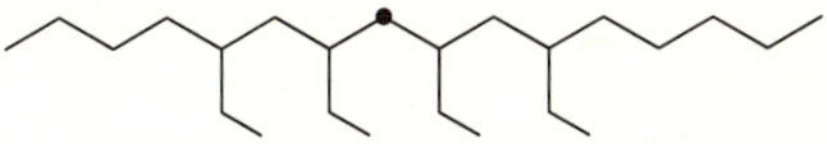
$I_2 = kN_{010}$	
$I_3 = 2k \sum_{i=1}^{i=n} N_{01(1)_i 10}$	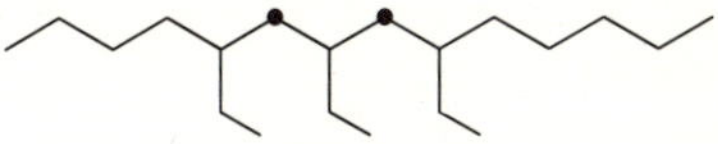
$I_4 = kN_{0110}$	
$I_5 = 2k \sum_{i=0}^{i=n} N_{01(1)_i 10}$	
$I_6 = k \sum_{i=0}^{i=n} iN_{01(1)_i 10}$	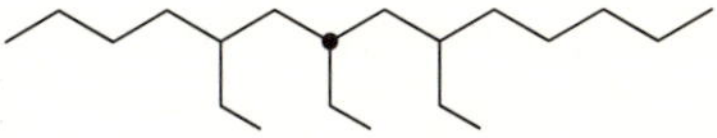
$I_7 = 2k \sum_{i=0}^{i=n} N_{01(1)_i 10}$	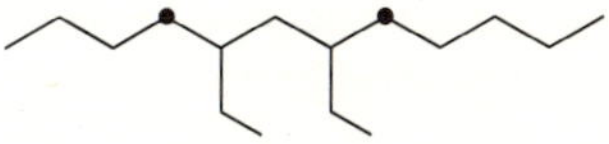
$I_8 = 2kN_{010}$	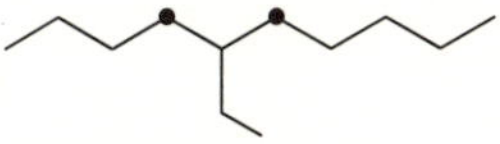
$I_9 = kN_{101}$	
$I_{10} = 2k \sum_{j=0}^{j=n} N_{10(0)_j 01}$	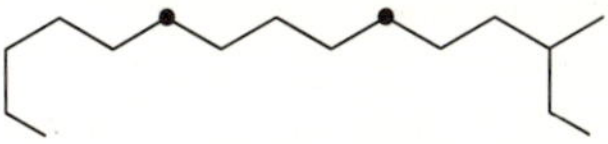
$I_{11} = k \sum_{j=0}^{j=n} (4j+3) N_{10(0)_j 01}$	
$I_{12} = 2kN_{010}$	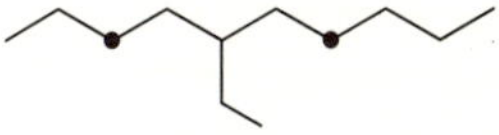
$I_{15} = kN_{010}$	
$I_{16} = kN_{010}$	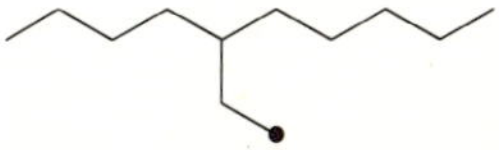

[a] From hydrogenated polybutadiene spectra relating peak intensity to sequence length (33).
[b] A black dot designates the specific carbon assignment. For those equations where summations are used to count the number of carbon atoms contributing to the resonance intensity, the structure shown is the first member of the series defined in the sum.

field. Assignments, according to a study by Clague *et al.* (32) and equations (33) describing each resonance intensity are given in Table 3-1. The chemical shift sensitivities vary from dyad to tetrad; however, it is possible to determine the relative concentration of each of the six unique triad combinations of 1,4 and 1,2 units.

The nomenclature used in Table 3-1 is as follows: 1,2 additions are represented by 1, 1,4 additions by 0, the number of 1,2 units in a run that are bonded only to other 1,2 units is given by i and analogously, the number of 1,4 units in a run that are bonded only to other 1,4 units is given by j. Table 3-1 also shows that some resonance intensities are described by different starting sequences than others. This part of the ^{13}C NMR analysis may be the most important and possibly the most difficult because the triad distribution and subsequent determinations of number-average sequence lengths depend upon correct descriptions of the relative intensities.

It may prove helpful to examine some of the descriptions of the NMR resonance intensities in more detail. The NMR chemical shift differences are governed by proximities to ethyl branches. A close inspection of the 101, 1001, and 10001 sequences shows that the center carbon of the 101 sequence gives rise to a unique ^{13}C NMR resonance I_9, because it is γ to both branched carbons, that is,

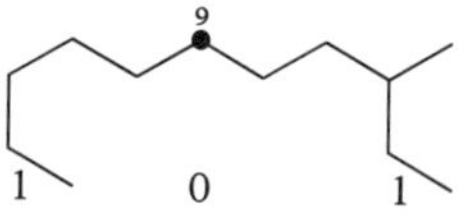

9

1 0 1

Carbons that are four or more removed from an ethyl branch give a separate resonance designated as I_{11}. The sequence 1001 contribute three carbons to I_{11}, and 10001 seven carbons:

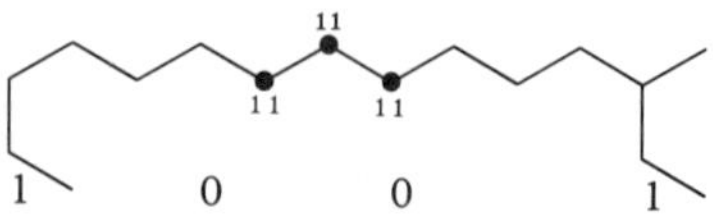

11

11 11

1 0 0 1

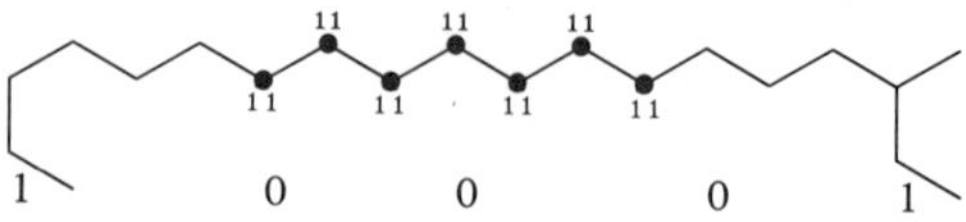

11 11 11

11 11 11 11

1 0 0 0 1

In general, the number of carbon atoms contributing to the intensity of resonance 11 for sequences of two 0 additions and higher is $(4j + 3)N_{10(0)_j01}$ because the insertion of each 1,4-butadiene after two 1,4 additions adds four more carbons that contribute to I_{11}. The equations describing the intensities of I_9 and I_{11}, therefore, are

$$I_9 = kN_{101} \tag{3.32}$$

$$I_{11} = k \sum_{j=0}^{j=n} (4j + 3)N_{10(0)_j01} \tag{3.33}$$

as given in Table 3-1. A similar line of reasoning can be used to derive the remaining equations in Table 3-1. Note that unique resonances are observed for the 010 and 0110 sequences and that the starting point for the summations in other cases vary from two to four consecutive 1 or 1,2 additions. For example, resonance 5 arises from terminal methine carbons from runs of 1,2 units two and longer. Only two carbons per run of 1,2 units contribute to I_5 regardless of the number of 1,2 or i units in an interior sequence, thus,

$$I_5 = 2k \sum_{i=0}^{i=n} N_{01(1)_i10} \tag{3.34}$$

and the intensity from resonance 5 can be used to count the number of sequences of 1,2 units in runs of two and longer. A combination of I_5 and I_8 (for 010) will count the total number of runs of 1,2 units per average chain. Any equation for the number-average sequence length for 1 additions will have either I_5 and I_8 or a descriptively similar combination in the denominator.

The six unique comonomeric triads are described as follows:

$$N_{111} = \sum_{i=0}^{i=n} iN_{01(1)_i10} = (1/k)I_6 \tag{3.35}$$

$$N_{110} = 2\sum_{i=0}^{i=n} N_{01(1)_i10} = (1/k)I_5 \tag{3.36}$$

$$N_{010} = N_{010} = (1/2k)I_8 \tag{3.37}$$

$$N_{101} = N_{101} = (1/k)I_9 \tag{3.38}$$

$$N_{001} = 2\sum_{j=0}^{j=n} N_{10(0)_j01} = (1/k)I_{10} \tag{3.39}$$

$$N_{000} = \sum_{j=0}^{j=n} jN_{10(0)_j01} = (1/4k)(I_{11} - \tfrac{3}{2}I_{10}) \tag{3.40}$$

Other combinations of peak intensities could be used to produce some of the relative triad concentrations. The above triad equations were selected because the required intensities were the least complicated by overlap with neighboring resonances. Equations (3.35)–(3.40) can be used to determine the relative triad concentrations because normalization eliminates the NMR constant k. Results, based on area measurements in ^{13}C spectra for each of the three hydrogenated polybutadienes, are given in Table 3-2. These data were also used to determine number-average sequence lengths of both 1,4 and 1,2 additions with Eqs (3.14) and (3.15). The results are given in Table 3-3.

1,2-Butadiene concentrations of 73, 47, and 25% obtained from the triad distribution using Eq. (3.25), agree closely with the reported percentages of 1,2-butadiene in each of the three hydrogenated polybutadienes in Tables 3-2 and 3-3.

It is apparent from the data in Tables 3-2 and 3-3 that ^{13}C NMR leads to a complete analysis of hydrogenated polybutadienes. This analysis could be extended to characterize polybutadienes by obtaining initially the *cis–trans* distribution (35) and then the carbon skeleton arrangement upon

TABLE 3-2 Comonomer Triad Distributions for Three Hydrogenated Polybutadienes[a]

Triad	Hydrogenated polybutadiene					
	72% 1,2[b]		50% 1,2[b]		26% 1,2[b]	
(111)	0.73	0.36 ± 0.001[c]	0.47	0.09 ± 0.011[c]	0.25	0.02 ± 0.005[c]
(110)		0.30 ± 0.005		0.23 ± 0.011		0.07 ± 0.002
(010)		0.07 ± 0.004		0.15 ± 0.007		0.16 ± 0.003
(101)	0.27	0.16 ± 0.006	0.53	0.16 ± 0.006	0.75	0.07 ± 0.004
(001)		0.10 ± 0.002		0.28 ± 0.011		0.32 ± 0.007
(000)		0.01 ± 0.003		0.09 ± 0.055		0.36 ± 0.013

[a] See Randall (33); (0 = 1,4-butadiene; 1 = 1,2-butadiene).
[b] Determined from IR analyses prior to hydrogenation (34).
[c] Standard deviations from areas by cutting and weighing.

TABLE 3-3 Number-Average Sequence Lengths for Both 1,4 and 1,2 Butadiene Additions[a]

Hydrogenated polybutadiene	$\bar{n}_0$	$\bar{n}_1$
72% 1,2	1.29 ± 0.02[b]	3.32 ± 0.10[b]
50% 1,2	1.77 ± 0.02	1.77 ± 0.05
26% 1,2	3.26 ± 0.12	1.28 ± 0.02

[a] In three hydrogenated polybutadienes (0 = 1,4; 1 = 1,2) (33).
[b] Standard deviation.

hydrogenation. Although not yet reported, direct ^{13}C NMR analysis of the polybutadiene terpolymer is probably feasible.

The triad distributions can also be used to examine the conformity of the monomer distribution (33) to either Markov or Bernoullian statistical models. Such analyses permit the monomer distribution to be identified as either random, ideally random, or dependent upon kinetic factors during polymerization. Because this is an important ingredient in the structural characterization of polymers, statistical analyses will be examined separately in Chapter 4.

3.4 MONOMER DISTRIBUTIONS AND NUMBER-AVERAGE SEQUENCE LENGTHS IN ETHYLENE–PROPYLENE COPOLYMERS

In the previous copolymer analysis for monomer distributions and number-average sequence lengths, three samples of hydrogenated polybutadienes were chosen because the monomer additions produced two structural units. The structures formed were unique and the polymer could be described by a succession of 0's and 1's. In this section, we shall discuss ethylene–propylene copolymers as an example of a second type of copolymer that requires a different approach to the structural analysis if a direct analysis is desired. The propylene monomer units can add in different directions, which create head-to-head or tail-to-tail monomer placements. Any disruption of normal head-to-tail (or tail-to-head) monomer additions is called inversion. Structurally, such copolymers should be described by a terpolymer terminology; however, one cannot identify all of the inverted monomer units. For example, the following ethylene–propylene additions lead to the same polymer structural entity; we shall use 0 for an ethylene unit, a 1 for a propylene unit where the methylene carbon is first, and 2 for a propylene unit where the methine carbon is first.

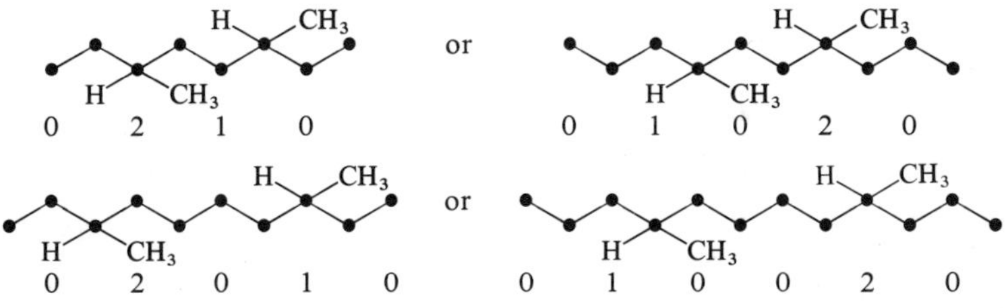

H CH$_3$ H CH$_3$
0 2 1 0 or 0 1 0 2 0

H CH$_3$ H CH$_3$
0 2 0 1 0 or 0 1 0 0 2 0

Further difficulties arise from the fact that a 1 2 addition is not the same as a 2 1 addition and likewise, a 1 0 2 is not the same as a 2 0 1,

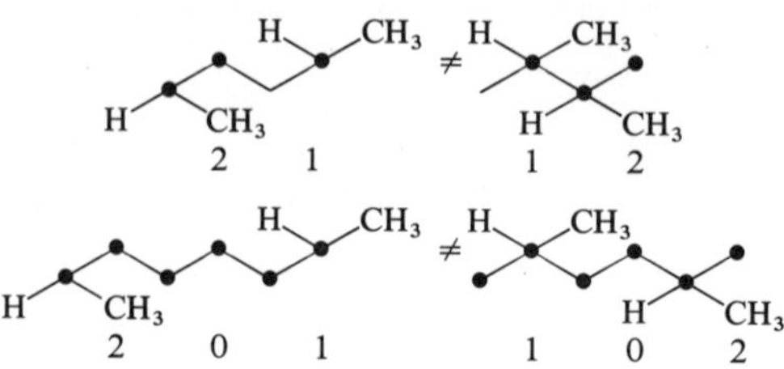

and, as illustrated, the 2 1 addition cannot be distinguished from the 1 0 2 addition. These problems are independent of the choice of vinyl monomer and will be a consideration in any copolymer of ethylene and a vinyl monomer.

In ^{13}C NMR spectra of ethylene–propylene copolymers, chemical shift differences occur because of the proximity and number of methyl branched carbon atoms. The ^{13}C NMR spectrum of a typical ethylene–propylene copolymer is shown in Fig. 3.4. Unique resonances are obtained according to the number of methylene carbons located between branches. The resonances in Fig. 3.4 are numbered consecutively from low to high field and the methylene carbon resonances are identified below according to the structural

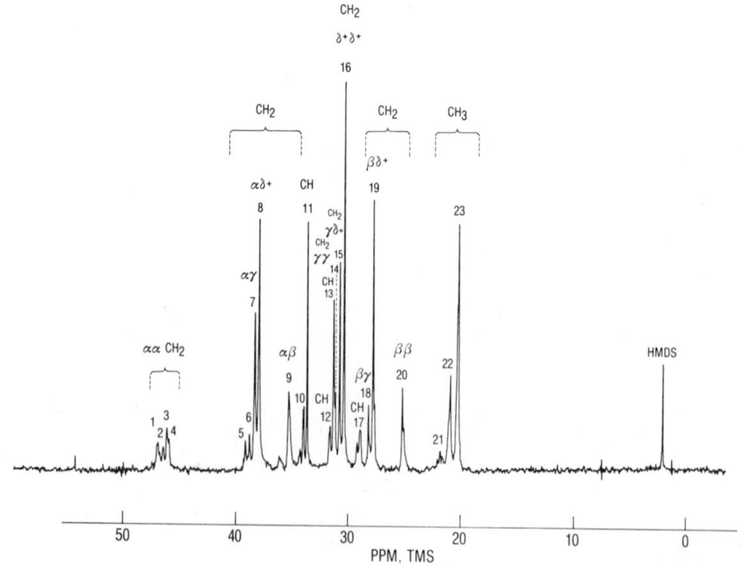

Fig. 3.4. Proton noise decoupled ^{13}C NMR spectrum at 25.2 MHz of a 40/60 ethylene–propylene copolymer in 1,2,4-trichlorobenzene at 120°C.

entity from which they originated (26). The Greek notations are those suggested by Carman (37).

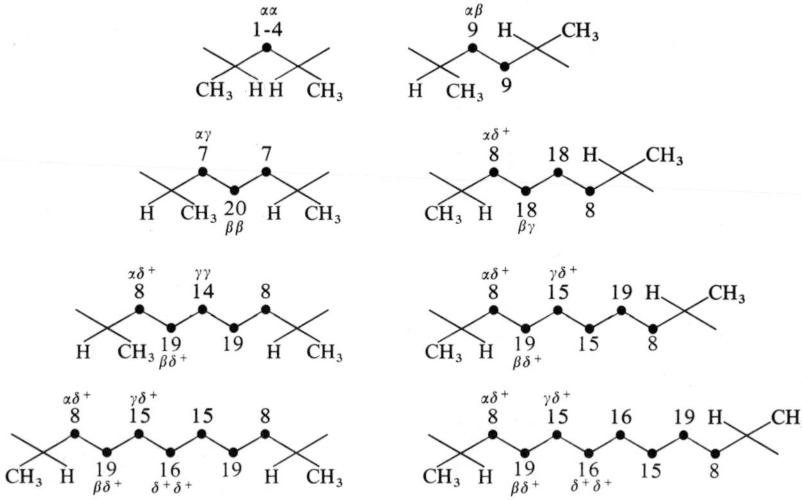

Note that unique resonances occur for methylene sequences in lengths of one to five. The same resonances (15 and 16, above) are observed for the central methylene carbons in sequences six and longer. The splittings for the $\alpha\alpha$ methylene carbon resonances (1–4, above) occur because of a combined tetrad and configurational sensitivity. Next nearest neighbor interactions, as well as near neighbor, determine these chemical shifts. In spite of these complicating factors, the area of the entire region designated as 1–4 is directly proportional to the $\alpha\alpha$ methylene carbon concentration.

Although it is customary to describe copolymer sequence lengths in terms of the monomers used in synthesis, it may be more meaningful for copolymers containing inverted monomer units to define them simply as a succession of methylene and methine carbons. Sufficient information is available from ^{13}C NMR to determine a number-average sequence length for runs of methylene carbons between methine carbons. Even though a description of ethylene–propylene copolymers containing inversions might for some purposes be desirable in terms of ethylene and propylene sequences (38), this description is precluded when ethylene sequences longer than two are present. The same methylene resonances are observed in sequences seven and longer independently of propylene inversion.

Number-average sequence lengths and the mole fraction of ethylene and propylene, however, can be easily obtained by describing the copolymer as a succession of 0's and 1's where 0 represents a methylene carbon and 1 represents a methine carbon. The following methylene ^{13}C intensities can be defined uniquely in terms of contributing carbon sequences:

$$I_{1\text{-}4} = kN_{101} \tag{3.41}$$

$$I_9 = 2kN_{1001} \tag{3.42}$$

$$I_{20} = kN_{10001} \tag{3.43}$$

$$I_{18} = 2kN_{100001} \tag{3.44}$$

$$I_{14} = kN_{1000001} \tag{3.45}$$

$$I_{15} = 2k \sum_{i=0}^{i=n} N_{1000(0)_i0001} \tag{3.46}$$

$$I_{16} = k \sum_{i=0}^{i=n} iN_{1000(0)_i0001} \tag{3.47}$$

where k is the NMR proportionality constant and $N_{10\cdots1}$ represents the number of $10\cdots1$ sequences per average polymer chain. Resonances 7 and 19 were not used in the above set because they are not unique. They also may describe

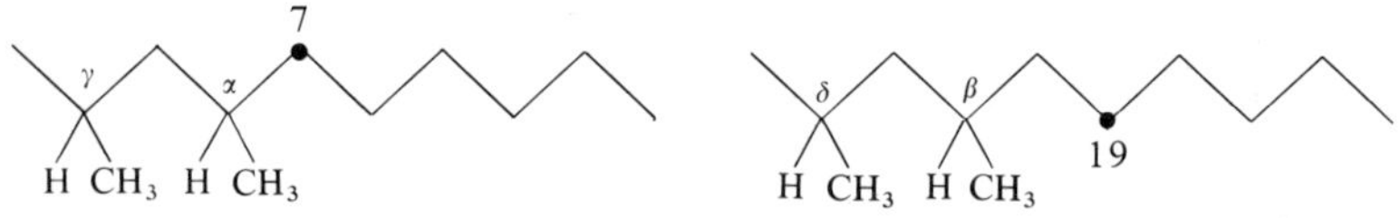

To determine the mole fraction of ethylene and propylene, one needs only the relative numbers of methylene and methine carbons. These are counted as follows:

$$N_0 = \sum_{i=0}^{i=n} iN_{1(0)_i1} \tag{3.48}$$

$$N_1 = \sum_{j=0}^{j=n} jN_{0(1)_j0} \tag{3.49}$$

where N_0 and N_1 are the numbers of methylene and methine carbons per average chain. In terms of the NMR intensities defined by Eqs. (3.41)–(3.47), Eq. (3.48) becomes

$$N_0 = (1/k)(I_{1\text{-}4} + I_9 + 3I_{20} + 2I_{18} + 5I_{14} + 3I_{15} + I_{16}) \tag{3.50}$$

and Eq. (3.49) is given by

$$N_1 = (1/k)(I_{CH_3}) \tag{3.51}$$

$$= (1/k)(I_{21} + I_{22} + I_{23}) \tag{3.52}$$

Note in Eq. (3.49) that N_{010} is the only term required because there is no evidence for 0110 sequences (see Fig. 1.7) while 01110 and higher sequences are not possible.

The mole fractions of ethylene and propylene are given by

$$(P) = 2N_1/(N_0 + N_1) \tag{3.53}$$

$$(E) = (N_0 - N_1)/(N_0 + N_1) \tag{3.54}$$

In addition to the ethylene–propylene monomer distribution, a number-average sequence length would also be desirable. It is possible to determine the number-average sequence length of uninterrupted methylene carbons but not the number-average sequence length of ethylene additions when inversion is present. The number-average sequence length of uninterrupted methylene carbons is given by

$$\bar{n}_0 = \sum_{i=0}^{i=n} iN_{1(0)_i1} \Big/ \sum_{i=1}^{i=n} N_{1(0)_i1} \tag{3.55}$$

which, upon expansion, is

$$\bar{n}_0 = \frac{N_{101} + 2N_{1001} + 3N_{10001} + 4N_{100001} + 5N_{1000001} + \sum_{i=0}^{i=n} (i + 6)N_{1000(0)_i0001}}{N_{101} + N_{1001} + N_{10001} + N_{100001} + N_{1000001} + \sum_{i=0}^{i=n} N_{1000(0)_i0001}} \tag{3.56}$$

In terms of the NMR intensities defined by Eqs. (3.41)–(3.47), Eq. (3.56) becomes

$$\bar{n}_0 = \frac{I_{1\text{-}4} + I_9 + 3I_{20} + 2I_{18} + 5I_{14} + 3I_{15} + I_{16}}{I_{1\text{-}4} + \frac{1}{2}I_9 + I_{20} + \frac{1}{2}I_{18} + I_{14} + \frac{1}{2}I_{15}} \tag{3.57}$$

An inspection of Eq. (3.56) also shows that the numerator simply counts the total number of methylene carbons and the denominator counts the total number of 10 sequences. Therefore, Eq. (3.55) can also be written as

$$\bar{n}_0 = \frac{2n_E + n_P}{n_P} \tag{3.58}$$

where n_E is the number of ethylene units per average chain and n_P is the corresponding number of propylene units. In terms of mole fractions, the number-average sequence length of uninterrupted methylene carbons is

$$\bar{n}_0 = 1 + 2(\mathrm{E})/(\mathrm{P}) \tag{3.59}$$

From Eq. (3.59) we conclude that the number-average sequence length of uninterrupted methylene carbons is only a reflection of the ethylene–propylene ratio contained in the copolymer and, therefore provides no additional insights into the desired distributions of methylene sequences. A more informative determination is available if one examines the number-average sequence length for methylene sequences two and longer, which is

$$\bar{n}_{2+} = \frac{I_9 + 3I_{20} + 2I_{18} + 5I_{14} + 3I_{15} + I_{16}}{\frac{1}{2}I_9 + I_{20} + \frac{1}{2}I_{18} + I_{14} + \frac{1}{2}I_{15}} \tag{3.60}$$

in terms of the NMR peak intensities defined by Eqs. (3.41)–(3.47).

The monomer distribution and number-average sequence lengths for the ethylene propylene copolymer shown in Fig. 3.2 are:

$$\bar{n}_0 = 3.5 \tag{3.61}$$

$$(n_0)_{\mathrm{cal'd}} = 3.0 \tag{3.62}$$

$$\bar{n}_{2+} = 4.6 \tag{3.63}$$

$$(\mathrm{P}) = 0.50 = 60\% \text{ by weight} \tag{3.64}$$

$$(\mathrm{E}) = 0.50 = 40\% \text{ by weight} \tag{3.65}$$

This comonomer distribution from ^{13}C NMR is in good agreement with the 40/60 ethylene-to-propylene ratio (by weight) used in the copolymer synthesis.

As stated previously, inversion of vinyl monomer units limits the amount of structure information directly available from ^{13}C NMR and will be a consideration in all copolymer analyses involving ethylene and vinyl monomers.[†] Without inversion, the analysis proceeds in exactly the same manner as for hydrogenated polybutadienes discussed in Section 3.3.

3.5 NUMBER-AVERAGE SEQUENCE LENGTHS AND DYAD DISTRIBUTIONS IN TERPOLYMERS

The mathematical derivations for number-average sequence lengths in terpolymers is more tedious than that for copolymers because a run of any

[†] See Section 6.16 where a statistical method for determining propylene inversion is discussed.

particular unit can be terminated by either of the other polymer units (39). In the terpolymer analysis, we shall use 0, 1, and 2 to designate the three types of unique backbone units that exist in the terpolymer. The number-average sequence length for 0 additions is given by

$$\bar{n}_0 = \frac{\sum_{i=0}^{i=n} iN_{1(0)_i1} + \sum_{i=0}^{i=n} iN_{1(0)_i2} + \sum_{i=0}^{i=n} iN_{2(0)_i2}}{\sum_{i=1}^{i=n} N_{1(0)_i1} + \sum_{i=1}^{i=n} N_{1(0)_i2} + \sum_{i=1}^{i=n} N_{2(0)_i2}} \tag{3.66}$$

Corresponding equations can be written for the 1 and 2 monomer units. The derivation for number-average sequence lengths follows the same lines of reasoning as the copolymer derivation. First, the dyad concentrations must be expressed in terms of run lengths, that is,

$$N_{00} = \sum_{i=0}^{i=n} iN_{10(0)_i1} + \sum_{i=0}^{i=n} iN_{10(0)_i2} + \sum_{i=0}^{i=n} iN_{20(0)_i2} \tag{3.67}$$

$$N_{11} = \sum_{i=0}^{i=n} iN_{01(1)_i0} + \sum_{i=0}^{i=n} iN_{01(1)_i2} + \sum_{i=0}^{i=n} iN_{21(1)_i2} \tag{3.68}$$

$$N_{22} = \sum_{i=0}^{i=n} iN_{02(2)_i0} + \sum_{i=0}^{i=n} iN_{02(2)_i1} + \sum_{i=0}^{i=n} iN_{12(2)_i1} \tag{3.69}$$

$$N_{01} = 2\sum_{i=0}^{i=n} N_{10(0)_i1} + \sum_{i=0}^{i=n} N_{10(0)_i2} \tag{3.70}$$

$$N_{02} = 2\sum_{i=0}^{i=n} N_{20(0)_i2} + \sum_{i=0}^{i=n} N_{20(0)_i1} \tag{3.71}$$

$$N_{12} = 2\sum_{i=0}^{i=n} N_{21(1)_i2} + \sum_{i=0}^{i=n} N_{21(1)_i0} \tag{3.72}$$

It is possible to express the heterogeneous dyad concentrations in more than one way. For example, N_{01} of Eq. (3.70) can also be expressed as

$$N_{01} = 2\sum_{i=0}^{i=n} N_{01(1)_i0} + \sum_{i=0}^{i=n} N_{01(1)_i2} \tag{3.73}$$

because the N_{01} dyads can be counted by summing over the total number of 1 runs terminated by at least one 0 or by summing over the number of 0 runs terminated by at least one 1. Dyads N_{02} and N_{12} can be handled in a similar manner, that is, by either of two possible expressions that depend upon the ease of extraction from NMR spectra. The chemical shift sensitivity and extent of overlap encountered usually determine the way in which the heterogeneous dyads are described.

The number-average sequence length for terpolymer units can be expressed as a function of dyad concentrations only by initially expanding the summations in Eq. (3.66) into the same form as Eqs. (3.67)–(3.72):

$$\bar{n}_0 = \frac{\sum_{i=0}^{i=n} (i+1)N_{10(0)_i 1} + \sum_{i=0}^{i=n} (i+1)N_{10(0)_i 2} + \sum_{i=0}^{i=n} (i+1)N_{20(0)_i 2}}{\sum_{i=0}^{i=n} N_{10(0)_i 1} + \sum_{i=0}^{i=n} N_{10(0)_i 2} + \sum_{i=0}^{i=n} N_{20(0)_i 2}} \tag{3.74}$$

then, substitution of Eqs. (3.67), (3.70), and (3.71) into Eq. (3.74) leads to

$$\bar{n}_0 = \frac{N_{00} + \frac{1}{2}N_{01} + \frac{1}{2}N_{02}}{\frac{1}{2}N_{01} + \frac{1}{2}N_{02}} \tag{3.75}$$

Equations for number-average sequence lengths for 1 and 2 units are developed in a similar manner. Final results are

$$\bar{n}_1 = \frac{N_{11} + \frac{1}{2}N_{01} + \frac{1}{2}N_{12}}{\frac{1}{2}N_{01} + \frac{1}{2}N_{12}} \tag{3.76}$$

$$\bar{n}_2 = \frac{N_{22} + \frac{1}{2}N_{12} + \frac{1}{2}N_{02}}{\frac{1}{2}N_{12} + \frac{1}{2}N_{02}} \tag{3.77}$$

As practiced in previous derivations, an example of a model terpolymer chain segment may prove instructive:

1 0 1 2 1 0 0 2 1 1 0 1 2 0 0 0 1 2 2 1 2 0 0 0 0 1 1 2 0 1 1 2 0 1 2 0 2 1

The number-average sequence length calculated with Eq. (3.74) is

$$\bar{n}_0 = \frac{5(1) + 1(2) + 1(3) + 1(4)}{5 + 1 + 1 + 1} = 1.75 \tag{3.78}$$

Conversely, there are six 00 dyads, nine 01 dyads, and seven 02 dyads, which by Eq. (3.75) again leads to a number-average sequence length of 1.75. The reader may wish to examine the number-average sequence lengths of the 1 and 2 units.

3.6 NUMBER-AVERAGE SEQUENCE LENGTHS AND TRIAD DISTRIBUTIONS IN TERPOLYMERS

There are 18 unique triads in terpolymers that increase substantially the quantitative information required for a complete analysis of number-average sequence lengths. These can be further divided into three groups of six that have common centers. Since the derivations would only be duplicated for the other units, it is necessary to examine only the number-average sequence length for 0 units and its six triads. Three of the 0-centered triads are unique, 101, 102, and 202. The remaining three are a function of run lengths:

$$N_{000} = \sum_{j=0}^{j=n} jN_{10(0)_j01} + \sum_{j=0}^{j=n} jN_{10(0)_j02} + \sum_{j=0}^{j=n} jN_{20(0)_j02} \tag{3.79}$$

$$N_{001} = 2\sum_{j=0}^{j=n} N_{10(0)_j01} + \sum_{j=0}^{j=n} N_{10(0)_j02} \tag{3.80}$$

$$N_{002} = 2\sum_{j=0}^{j=n} N_{20(0)_j02} + \sum_{j=0}^{j=n} N_{10(0)_j02} \tag{3.81}$$

After expanding the summations of Eq. (3.66) into the form of Eqs. (3.79)–(3.81), the number-average sequence length for 0 additions is given by

$$\bar{n}_0 = \frac{\begin{array}{l} N_{101} + \sum_{j=0}^{j=n} (j+2)N_{10(0)_j01} + N_{102} \\ \quad + \sum_{j=0}^{j=n} (j+2)N_{10(0)_j02} + N_{202} + \sum_{j=0}^{j=n} (j+2)N_{20(0)_j02} \end{array}}{\begin{array}{l} N_{101} + \sum_{j=0}^{j=n} N_{10(0)_j01} + N_{102} \\ \quad + \sum_{j=0}^{j=n} N_{10(0)_j02} + N_{202} + \sum_{j=0}^{j=n} N_{20(0)_j02} \end{array}} \tag{3.82}$$

As a function of triad concentrations defined by Eqs. (3.79)–(3.81), the number-average sequence length of 0 units becomes

$$\bar{n}_0 = \frac{N_{101} + N_{102} + N_{202} + N_{002} + N_{001} + N_{000}}{N_{101} + N_{102} + N_{202} + \frac{1}{2}N_{002} + \frac{1}{2}N_{001}} \tag{3.83}$$

Equation (3.83) can also be derived intuitively because the number-average sequence length is given by the total number of units of a specific type divided by the number of runs containing that unit. The number of 0 units is the sum of the six unique triad combinations with 0 as the center unit. The number of runs is given by the three triads N_{101}, N_{102}, and N_{202} plus the number of sequences two and longer which is simply $\frac{1}{2}N_{001} + \frac{1}{2}N_{002}$. For any particular run two or longer, there are either two 001 or 002 triads per run or both a 001 and a 002 triad.

The corresponding equations for the number-average sequence length for the 1 and 2 terpolymer units have the same form as the 0 case and can be derived similarly. They are

$$\bar{n}_1 = \frac{N_{010} + N_{012} + N_{212} + N_{111} + N_{110} + N_{112}}{N_{010} + N_{012} + N_{212} + \frac{1}{2}N_{110} + \frac{1}{2}N_{112}} \tag{3.84}$$

$$\bar{n}_2 = \frac{N_{020} + N_{021} + N_{121} + N_{222} + N_{220} + N_{221}}{N_{020} + N_{021} + N_{121} + \frac{1}{2}N_{220} + \frac{1}{2}N_{221}} \tag{3.85}$$

In the model chain segment of a 1,2,0 terpolymer discussed previously, there are two 101 triads, two 201, one 202, three 200, three 100, and three 000 triads. Substitution into Eq. (3.83) gives a number-average sequence length for 0 additions of 1.75 in agreement with the result calculated from the dyad distribution.

Equations (3.83)–(3.85) also reduce to the dyad equations (3.75)–(3.77) after substitution with the appropriate triad–dyad relationships:

$$N_{00} = N_{000} + \tfrac{1}{2}N_{001} + \tfrac{1}{2}N_{002} \tag{3.86}$$

$$N_{11} = N_{111} + \tfrac{1}{2}N_{110} + \tfrac{1}{2}N_{112} \tag{3.87}$$

$$N_{22} = N_{222} + \tfrac{1}{2}N_{220} + \tfrac{1}{2}N_{221} \tag{3.88}$$

$$N_{01} = 2N_{010} + N_{110} + N_{210} = 2N_{101} + N_{001} + N_{201} \tag{3.89}$$

$$N_{02} = 2N_{020} + N_{220} + N_{120} = 2N_{202} + N_{002} + N_{102} \tag{3.90}$$

$$N_{12} = 2N_{121} + N_{221} + N_{021} = 2N_{212} + N_{112} + N_{012} \tag{3.91}$$

Proofs of the triad–dyad relationships and the alternate heterogeneous dyad definitions can be obtained in a manner analogous to the copolymer derivation. These relationships are useful in terpolymer analyses of number-average sequence lengths when complex ^{13}C NMR spectra are encountered or overlap presents problems. An example of the method developed in this section is given in the following discussion of hydrogenated butadiene–styrene copolymers.

3.7 NUMBER-AVERAGE SEQUENCE LENGTHS IN HYDROGENATED BUTADIENE–STYRENE COPOLYMERS

Carbon-13 NMR spectra of terpolymers are typically complex because monomer units may not exhibit the same chemical shift sensitivities and overlap among key resonances may exclude information critical to analyses of sequence distributions. An example of a successful terpolymer analysis is found in a ^{13}C NMR study of hydrogenated butadiene–styrene copolymers. Although a copolymer by definition, hydrogenated butadiene–styrene copolymers consist of three different structural units: 1,4- and 1,2-butadiene additions and styrene. As discussed in Section 3.3 in a corresponding analysis of hydrogenated polybutadienes, all six butadiene triads could be identified and measured quantitatively. In the terpolymer analysis, a complete set of 18 triads or six dyads consisting of 1,4- and 1,2-butadiene and styrene units is necessary before number-average sequence lengths can be determined. In many cases the success of such analyses will depend upon the intrinsic NMR chemical shift sensitivities and the ability of the analyst to obtain unambiguous chemical shift assignments.

Carbon-13 NMR spectra of four hydrogenated butadiene–styrene copolymers, which differ in composition and sequence distributions, are shown in Figs. 3.5–3.8. The resonances are labeled from 1 to 26 from low to high field in order of appearance although all of these resonances are not present in any one spectrum in Figs. 3.5–3.8. Assignments, summarized in Table 3-4, were based on the Grant and Paul rules and structural trends established among these copolymer samples (39). The spectra vary in complexity depending primarily upon arrangements and distributions of styrene units. Polymer A contained only isolated styrene additions, polymer B contained principally isolated styrene units but also had low concentrations of blocked styrene units, polymer C was composed of a mixture of blocked and isolated

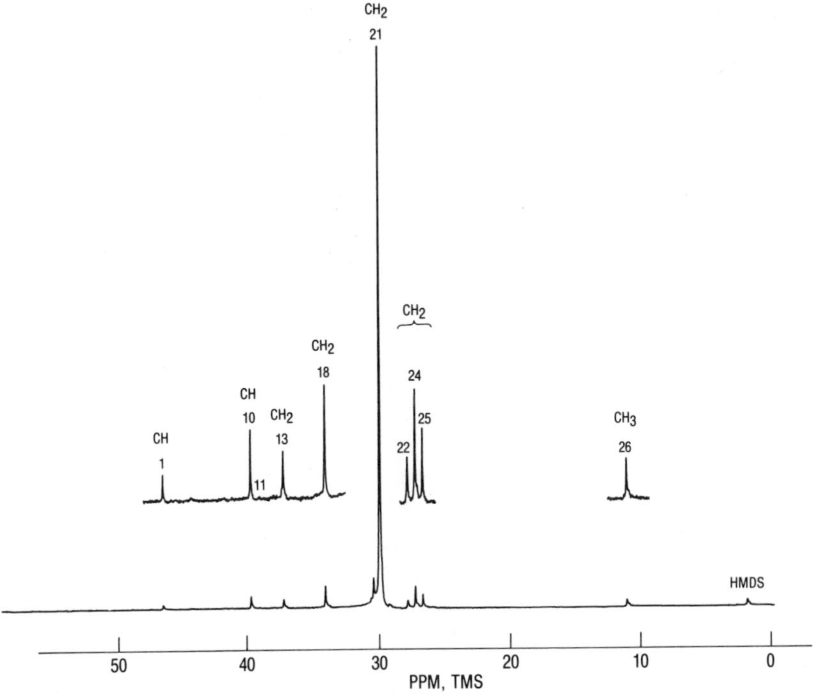

Fig. 3.5. Proton noise decoupled ^{13}C NMR spectrum at 25.2 MHz of a hydrogenated butadiene–styrene copolymer (4 mole % styrene, 6 mole % 1,2-butadiene, and 90 mole % 1,4-butadiene) in 1,2,4-trichlorobenzene at 120°C. (Co polymer A.)

styrene additions, and polymer D, which was high in 1,2-butadiene additions, contained predominantly isolated styrene additions.

In Table 3-4, equations relating NMR peak intensities to sequence lengths can be used to obtain the six unique dyad concentrations:

$$N_{00} = \tfrac{1}{4}(I_{21} + \tfrac{1}{2}I_{20})/k \tag{3.92}$$

$$N_{01} = (I_{18} + I_{12})/k \tag{3.93}$$

$$N_{02} = (2I_{1\text{-}3} + I_4 + I_5)/k \tag{3.94}$$

$$N_{22} = (I_7 + \tfrac{1}{2}I_4 + \tfrac{1}{2}I_8)/k \tag{3.95}$$

$$N_{12} = (2I_{8s} + I_5 + I_8)/k \tag{3.96}$$

$$N_{11} = (I_{16\text{-}17} - I_{12} + \tfrac{1}{2}I_{15} + \tfrac{1}{2}I_{10} + I_{11} - \tfrac{1}{4}I_{18})/k \tag{3.97}$$

These equations may be derived by using the dyad–triad necessary relationships defined by Eqs. (3.86)–(3.91) and the peak intensity equations in Table 3-4.

After obtaining area measurements through curve resolving, we may calculate the dyad distributions with Eqs. (3.92)–(3.97). Final results are

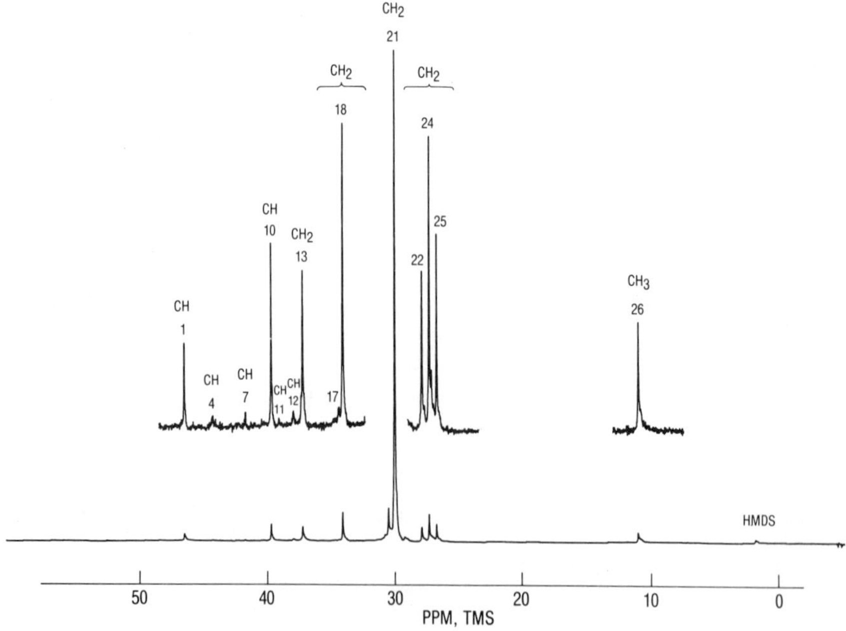

Fig. 3.6. Proton noise decoupled ^{13}C NMR spectrum at 25.2 MHz of a hydrogenated butadiene–styrene copolymer (6 mole % styrene, 10 mole % 1,2-butadiene, and 84 mole % 1,4 butadiene) in 1,2,4-trichlorobenzene at 120°C. (Copolymer B).

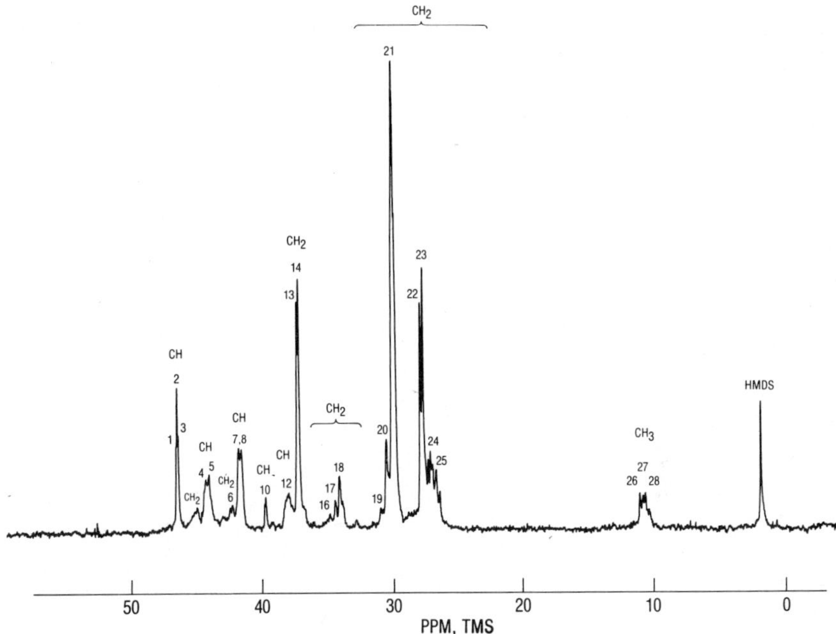

Fig. 3.7. Proton noise decoupled ^{13}C NMR spectrum at 25.2 MHz of a hydrogenated butadiene–styrene copolymer (43 mole % styrene, 15 mole % 1,2-butadiene, and 42 mole % 1,4-butadiene) in 1,2,4-trichlorobenzene at 120°C. (Copolymer C.)

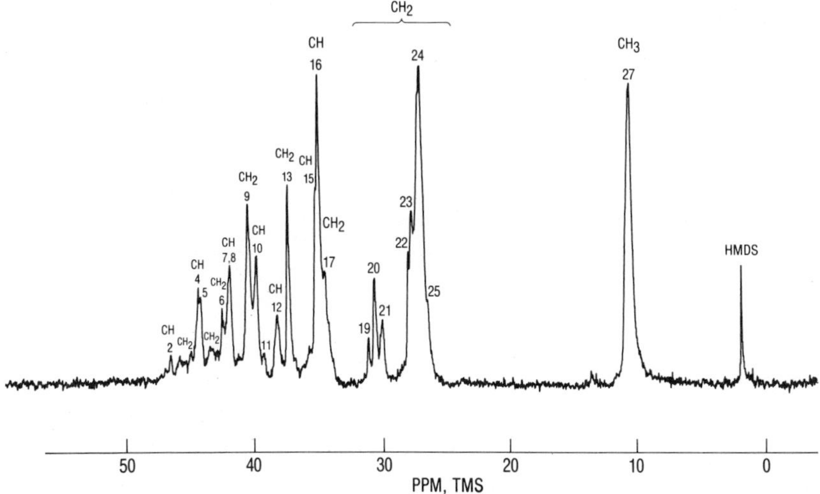

Fig. 3.8. Proton noise decoupled ^{13}C NMR spectrum at 25.2 MHz of a hydrogenated butadiene–styrene copolymer (29 mole % styrene, 55 mole % 1,2-butadiene, and 15 mole % 1,4-butadiene) in 1,2,4-trichlorobenzene at 120°C. (Copolymer D.)

TABLE 3-4 Equations for Resonance Areas 1–26[a]

Equation	Assignment[b,c]
$I_{1\text{-}3} = kN_{020}$	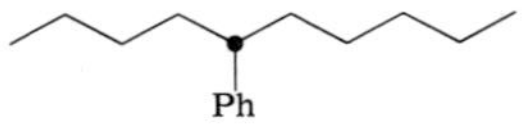 Ph
$I_4 = 2k \sum_{i=2}^{i=n} N_{0(2)_i0} + k \sum_{i=2}^{i=n} N_{0(2)_i1}$	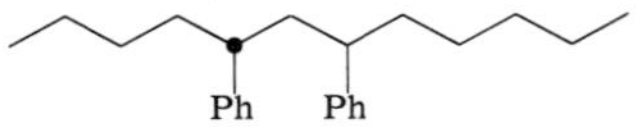 Ph Ph
$I_5 = kN_{021}$	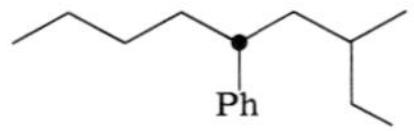 Ph
$I_6 = kN_{21}$	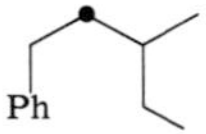 Ph
$I_7 = k \sum_{i=0}^{i=n} iN_{02(2)_i20} + k \sum_{i=0}^{i=n} iN_{02(2)_i21} + k \sum_{i=0}^{i=n} iN_{12(2)_i21}$	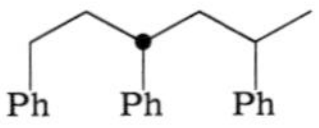 Ph Ph Ph
$I_8 = 2k \sum_{i=2}^{i=n} N_{1(2)_i1} + k \sum_{i=2}^{i=n} N_{1(2)_i0}$	 Ph Ph (Ph)[d]
$I_9 = k \sum_{i=1}^{i=n} (i-1)N_{1(1)_i1}$	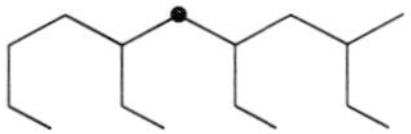
$I_{10} = kN_{010} + 2k \sum_{i=3}^{i=n} N_{0(1)_i0} + k \sum_{i=2}^{i=n} N_{0(1)_i2}$	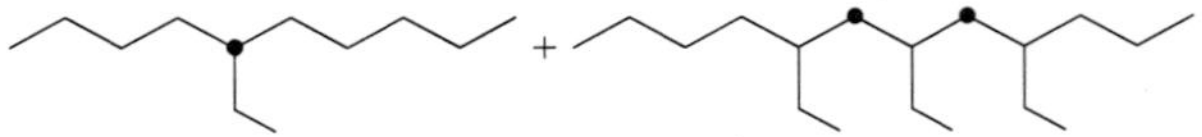 +
$I_{11} = kN_{0110}$	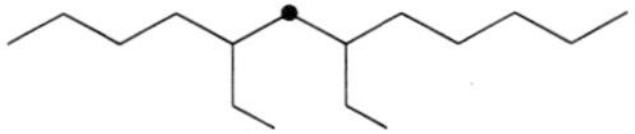
$I_{12} = 2k \sum_{i=3}^{i=n} N_{0(1)_i0} + k \sum_{i=1}^{i=n} N_{0(1)_i2}$	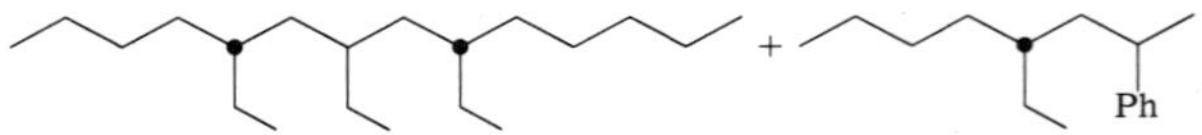 + Ph
$I_{13\text{-}14} = 2k \sum_{i=1}^{i=n} N_{0(2)_i0} + k \sum_{i=1}^{i=n} N_{0(2)_i1}$	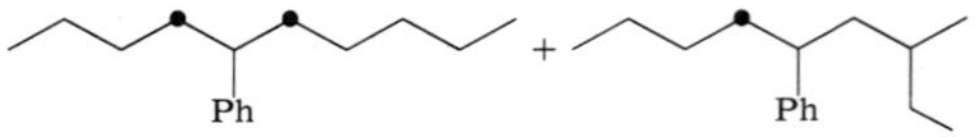 Ph + Ph

$$I_{15} = 2k \sum_{i=2}^{i=n} N_{2(1)_i2} + k \sum_{i=2}^{i=n} N_{2(1)_i0}$$

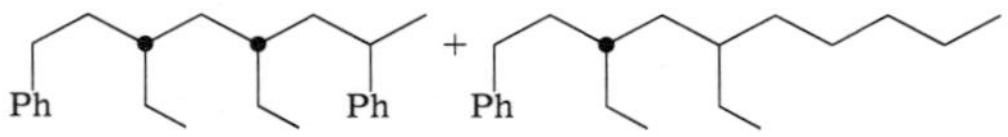

$$I_{16\text{-}17} = k \sum_{i=0}^{i=n} iN_{01(1)_i10} + k \sum_{i=0}^{i=n} iN_{01(1)_i12} + k \sum_{i=0}^{i=n} iN_{21(1)_i12} + 2k \sum_{i=2}^{i=n} N_{0(1)_i0} + k \sum_{i=1}^{i=n} N_{0(1)_i2}$$

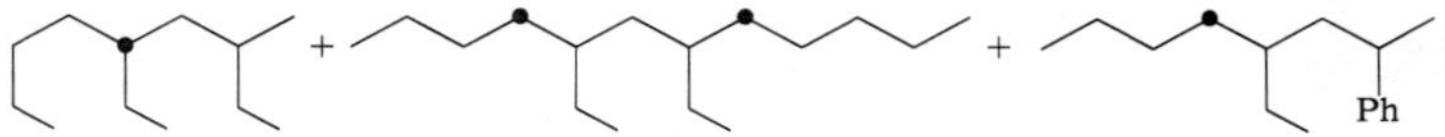

$$I_{18} = 2kN_{010}$$

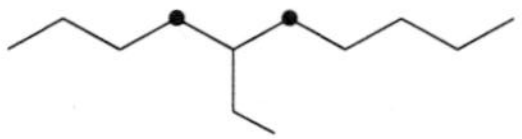

$$I_{19} = kN_{101} + kN_{102} + kN_{202}$$

$$I_{20} = 2k \sum_{i=2}^{i=n} N_{R(0)_iR}$$

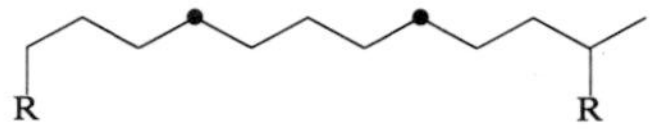

$$I_{21} = k \sum_{i=0}^{i=n} (4i + 3)N_{R0(0)_i0R}$$

$$I_{22\text{-}23} = 2k \sum_{i=1}^{i=n} N_{0(2)_i0} + k \sum_{i=1}^{i=n} N_{0(2)_i1}$$

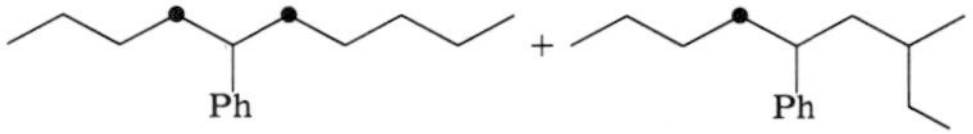

$$I_{24} = 2k \sum_{i=1}^{i=n} N_{0(1)_i0} + k \sum_{i=1}^{i=n} N_{0(1)_i2}$$

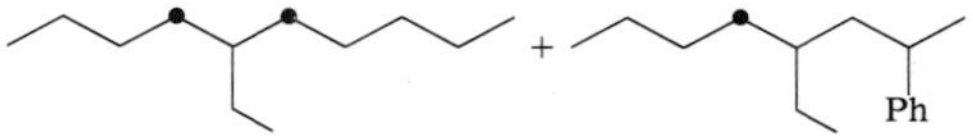

$$I_{25} = kN_{010}$$

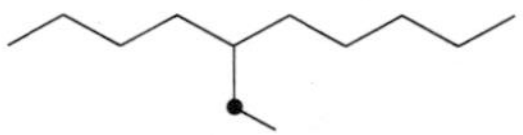

$$I_{26} = kN_{010}$$

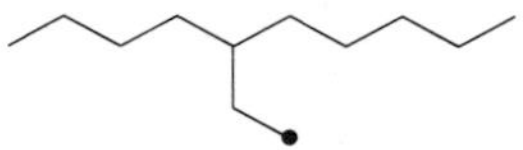

[a] Relating peak intensity to sequence length, and assignments for hydrogenated butadiene–styrene copolymers A, B, C, and D (39). 0 = 1,4-butadiene; 1 = 1,2-butadiene; 2 = styrene.

[b] Heavy black dot denotes the specific carbon assigned, for those equations where summations are used to count the number of carbon atoms contributing to the resonance intensity, the structure shown is the first member of the series defined in the sum.

[c] R = ethyl or phenyl. [d] Has appeared as a shoulder (8s) in the ^{13}C spectrum of polymer D.

given in Table 3-5. Number-average sequence lengths determined from the dyad values in Table 3-5 and calculated with Eqs. (3.75)–(3.77) are given in Table 3-6. The comonomer distribution, which is valuable for comparison with results from other techniques, can be calculated from the dyad distribution. The necessary dyad–monomer content relationships are

$$(0) = (00) + \tfrac{1}{2}(01) + \tfrac{1}{2}(02) \tag{3.98}$$

$$(1) = (11) + \tfrac{1}{2}(01) + \tfrac{1}{2}(12) \tag{3.99}$$

$$(2) = (22) + \tfrac{1}{2}(02) + \tfrac{1}{2}(12) \tag{3.100}$$

Final comonomer distributions, obtained from Eqs. (3.98) to (3.100) with the dyad values in Table 3-5, are given in Table 3-7 for the hydrogenated butadiene–styrene copolymers. A corresponding measurement obtained from ^{1}H NMR data is included (39).

Whenever possible, it is a good practice in ^{13}C analyses of comonomer distributions and number-average sequence lengths to obtain data for an independent comparison. Satisfactory agreement between ^{1}H and ^{13}C NMR results in Table 3-7 lends support to the ^{13}C NMR results for the higher

TABLE 3-5 Monomer Dyad Distributions for Hydrogenated Butadiene–Styrene Copolymers[a]

Sample[b]	(00)	(01)	(02)	(22)	(12)	(11)
A	0.80	0.12	0.08	0.00	0.00	0.00
B	0.70	0.17	0.10	0.01	0.00	0.01
C	0.17	0.13	0.36	0.16	0.17	0.00
D	0.02	0.11	0.15	0.13	0.17	0.41

[a] 0 = 1,4-butadiene; 1 = 1,2-butadiene; and 2 = styrene (39).

[b] Sample A contains isolated styrene units only, sample B contains principally isolated styrene units, sample C contains both blocked and isolated styrene units, and sample D contains predominantly isolated styrene units but is high in 1,2-butadiene additions.

TABLE 3-6 Number-Average Sequence Lengths for 1,4-Butadiene(0), 1,2-Butadiene(1), and Styrene(2) Additions in Hydrogenated Butadiene–Styrene Copolymers[a]

Sample	$\bar{n}_0$	$\bar{n}_1$	$\bar{n}_2$
A	8.0	1.0	1.0
B	5.9	1.0_5	1.0_6
C	1.9	1.0	1.6
D	1.2	3.9	1.8

[a] See Randall (39).

TABLE 3-7 Monomer Distributions for Hydrogenated Butadiene–Styrene Copolymers[a]

	Polymer A		Polymer B		Polymer C		Polymer D	
Monomer	^{13}C	^{1}H	^{13}C	^{1}H	^{13}C	^{1}H	^{13}C	^{1}H
Styrene	4	6	6	6	43	42	29	25
1,4-Bd	90	89	84	89	42	36	15	23
1,2-Bd	6	5	10	5	15	22	55	52

[a] Obtained from both ^{13}C and ^{1}H NMR data (39).

comonomer distributions. The procedure discussed in this section can be applied to any terpolymer analysis involving vinyl monomers and ethylene where inversions have not occurred. The problem of inversion may lead to a limited analysis and should be considered on an individual copolymer basis. Head-to-head and tail-to-tail styrene additions can be easily distinguished from those involving butadiene and were not a consideration in the hydrogenated butadiene–styrene terpolymer analysis. Inversions of either 1,2-butadiene or styrene units are not likely from a standpoint of the mechanism proposed (40) for alkyl lithium-initiated butadiene polymerizations.

4 Statistical Analyses of Monomer Distributions and Number-Average Sequence Lengths

A technique frequently practiced when making ^{13}C NMR chemical shift assignments is the inspection of observed comonomer distributions for conformity to a statistical model. It will be the purpose of this chapter to introduce the reader to statistical analyses by examining initially Bernoullian and then first-order Markov statistical models for the more practical aspects of polymer applications, that is, NMR chemical shift assignments and subsequent determinations of comonomer distributions and number-average sequence lengths. Bovey (9) and Price (6) pioneered the development of statistical analyses of polymers and excellent theoretical treatises are available on the subjects of Bernoullian and Markovian statistics (4, Chapter II; 41, Chapter 7).

The most frequently used statistical models are Bernoullian and first-order Markov. The statistical frameworks of both these models are developed in terms of probabilities for specific monomer additions (42). Bernoullian models describe a random distribution of monomer units where the probability of a monomer addition is independent of the outcome of any previous addition. All first- (and higher-) order Markovian models describe those circumstances where a probability of addition is dependent upon previous events: a first-order Markov model depends upon a single preceding event; a second-order Markov model depends upon the outcome of two previous events, and so on. Comonomer distributions are then defined by probability expressions for complete sets of mole fractions, that is, normalized dyad or triad, etc., concentrations. Rules, which allow testing of an observed comonomer distribution for conformity with a specified model, can be developed from the mathematical framework.

Bernoullian models may also be called zero-order Markov because there is no dependency upon preceding events. We shall examine both Bernoullian and first-order Markov models and applications to polymer analyses. The more practical aspects of these applications will be discussed because pitfalls are present if the analyses are not handled both rigorously and cautiously.

4.1 BERNOULLIAN STATISTICAL ANALYSES

The Bernoullian statistical model is the most frequently examined probability model in polymer analyses. Because it does describe a perfectly random distribution of two or more types of monomer additions, a Bernoullian distribution is a common occurrence in vinyl homopolymers as well as some copolymers. Bernoullian behavior has been observed for *cis–trans* distributions among 1,4 additions in polybutadienes (43–45) for the comonomer distribution in ethylene–vinyl acetate copolymers (20) and for configurational distributions in polymers such as polystyrene (19), poly(vinyl chloride) (46) and poly(vinyl alcohol) (47). Most often, the Bernoullian model has been used for testing configurational assignments in ^{13}C NMR spectra; although, as will be seen later, adherence to any particular probability scheme leads to a complete structural characterization. The Bernoullian model can be tested more rigorously than other statistical models because the mathematical framework requires only one independent variable; thus identifications of conformity to Bernoullian behavior can be made with confidence.

Let us now examine the mathematical framework of the Bernoullian copolymer model as developed by Price (6). As in Chapters 2 and 3, 0 is used to designate one comonomer unit and 1, the other. A polymer chain is then defined as a succession of individual states and probabilities are defined for a transition from one state to the succeeding state. In Bernoullian analyses of copolymers, a state is defined as a single polymer unit. The associated transition probabilities† are then defined:

Initial state	Add	Final state	Transition probability
0 or 1	0	0	P_0
	1	1	P_1

It is unnecessary to designate specific initial states because the transition probabilities in Bernoullian systems are independent of the preceding state. Transition probabilities P_0 and P_1 consequently are defined only by the final states. The number of transitions are also limited to either 0 or 1 additions, thus,

$$P_0 + P_1 = 1 \tag{4.1}$$

and only one independent variable is required for a Bernoullian analysis. These transition probabilities ultimately represent the probability of finding either 0 or 1 units; therefore they correspond to the 0 and 1 mole fractions observed for the copolymer. The simplicity of the Bernoullian model is found in this result because a comonomer mole fraction is all that is needed to test for conformity to Bernoullian behavior. Examples of Bernoullian descriptions of the relative concentrations of various comonomer sequences are

Sequence:	0	1	01	111	etc.
Mole fraction:	P_0	$1 - P_0$	$P_0(1 - P_0)$	$(1 - P_0)^3$	

A complete characterization of a copolymer is possible if conformity to Bernoullian behavior can be unambiguously established. One of the most appropriate tests for Bernoullian behavior is to compare complete sets of experimentally observed comonomer mole fractions with those calculated from individual unit mole fractions. For copolymers, the analysis is usually limited to dyad and triad distributions because of the availability of experimental data, typically from NMR.

As seen in these examples, development of a complete set of Bernoullian probability expressions for a comonomer distribution of any particular length can be accomplished by inspection. One must determine only the total number of sequences and degeneracies, as shown for dyad and triad distributions:

$$(00) = P_0^2 \tag{4.2}$$

$$(10) + (01) = 2P_0(1 - P_0) \tag{4.3}$$

$$(11) = (1 - P_0)^2 \tag{4.4}$$

† These are also called conditional probabilities.

and

$$(000) = P_0^{\,3} \tag{4.5}$$

$$(100) + (001) = 2P_0^{\,2}(1 - P_0) \tag{4.6}$$

$$(101) = P_0(1 - P_0)^2 \tag{4.7}$$

$$(010) = P_0^{\,2}(1 - P_0) \tag{4.8}$$

$$(110) + (011) = 2P_0(1 - P_0)^2 \tag{4.9}$$

$$(111) = (1 - P_0)^3 \tag{4.10}$$

A Bernoullian fit is usually established after a comparison of an observed dyad or triad distribution with that calculated with Eqs. (4.2)–(4.10). For some polymers, the individual unit mole fractions may not be known. It is then customary to iterate over values of P_0 until the best solution is obtained.

Other useful relationships, which are independent of P_0, are also available for detecting Bernoullian systems. For the triad case,

$$[(100) + (001)]/(010) = [(011) + (110)]/(101) = 2.0 \tag{4.11}$$

Note that as a natural consequence of Bernoullian behavior it is the number of given units within a sequence and not the arrangement that determines the relative concentrations. Likewise, for dyads,

$$[(10) + (01)]/(00)^{1/2}(11)^{1/2} = 2.0 \tag{4.12}$$

When examining an observed system for conformity to Bernoullian statistics, one should use a dyad distribution cautiously because only two independent observations are available for testing one independent variable. A triad distribution is more suitable because five independent observations are available for fitting a Bernoullian model. After a fit has been established, the transition probability can be used to calculate the mole fraction of any particular sequence of monomer units.

In addition to determining sequence concentrations, the number-average sequence lengths of successive 0 or 1 additions can be calculated by substituting the appropriate probability expressions into any of the equations derived in Chapter 3 that relate monomer distributions to number-average sequence lengths. For example,

$$\bar{n}_0 = \frac{(00) + \frac{1}{2}(01)}{\frac{1}{2}(01)} = \frac{1}{1 - P_0} \tag{4.13}$$

and

$$\bar{n}_1 = \frac{(11) + \frac{1}{2}(01)}{\frac{1}{2}(01)} = \frac{1}{P_0} \tag{4.14}$$

Before proceeding further with a discussion of the criteria necessary for distinguishing Bernoullian from possible first-order Markov fits, it may be instructive to examine the first-order Markov model, the differences expected, and finally an application to a system where conformity to a specific statistical behavior is observed.

4.2 FIRST-ORDER MARKOV STATISTICAL ANALYSES OF COPOLYMERS

Polymeric systems that conform to first-order and higher-Markovian statistical models have probabilities of addition that are dependent upon the outcome of one or more previous events. We shall consider the first-order Markov model in detail beginning with the Price definition of initial and final states and the possible transitions that take place. A first-order Markov system has transition probabilities that depend upon the outcome of the immediately preceding event, that is,

Initial state	Add	Final state	Transition probability
0	0	0	P_{00}
0	1	1	P_{01}
1	0	0	P_{10}
1	1	1	P_{11}

The four transition probabilities are related as follows:

$$P_{00} + P_{01} = 1 \tag{4.15}$$

$$P_{11} + P_{10} = 1 \tag{4.16}$$

because a chain ending in 0 can only add a zero or one and likewise, a chain ending in 1 has only this prerogative. Thus only two transition probabilities or independent variables are required to define a first-order Markov process. Before a sequence can be defined in terms of these transition probabilities, we must also define the probability of finding a particular initial state. These probabilities, descriptively, are given by ${}^1P_0$ and ${}^1P_1$. Because a given unit must be either 0 or 1, the total probability is unity, that is,

$${}^1P_0 + {}^1P_1 = 1 \tag{4.17}$$

and

$${}^1P_0 P_{00} + {}^1P_1 P_{10} = {}^1P_0 \tag{4.18}$$

Equation (4.18) defines the probability for finding a 0 unit and follows from the consideration that the preceding unit must be either 0 or 1. Likewise, for 1 units,

$$^1P_1 P_{11} + {}^1P_0 P_{01} = {}^1P_1 \tag{4.19}$$

Substitution of Eqs. (4.15)–(4.17) into (4.18) and (4.19) leads to expressions which state the probability of finding 0 and 1 units in terms of transition probabilities only, that is,

$$^1P_0 = P_{10}/(P_{01} + P_{10}) \tag{4.20}$$

$$^1P_1 = P_{01}/(P_{01} + P_{10}) \tag{4.21}$$

Any chain segment can now be defined by expressions containing just two first-order Markov transition probabilities, for example,

Sequence:	0	01	111
Mole fraction:	$\frac{P_{10}}{P_{01} + P_{10}}$	$\frac{P_{10}P_{01}}{P_{01} + P_{10}}$	$\frac{P_{01}(1 - P_{10})^2}{P_{01} + P_{10}}$

Thus, any specific comonomer distribution can be described for fitting with an observed distribution.

One of the problems associated with statistical analyses is the availability of sufficient observations to establish conformity to a particular model without equivocation. A minimum requirement for analyses establishing first-order Markovian behavior is that there must be enough experimental observations to lead to an overdetermined analysis. A dyad distribution of monomer units, therefore, cannot be used to test for conformity to Markovian statistics because any solution for probabilities is exact. Only two independent observations are available for determining two unknowns:

$$(00) = P_{10}(1 - P_{01})/(P_{01} + P_{10}) \tag{4.22}$$

$$(10) + (01) = 2P_{10}P_{01}/(P_{01} + P_{10}) \tag{4.23}$$

$$(11) = P_{01}(1 - P_{10})/(P_{01} + P_{10}) \tag{4.24}$$

A triad distribution, however, has an adequate number of degrees of freedom for a first-order Markovian analysis. The triad equations are given in terms of the probabilities P_{01} and P_{10}:

$$(000) = P_{10}(1 - P_{01})^2/(P_{01} + P_{10}) \tag{4.25}$$

$$(100) + (001) = 2P_{01}P_{10}(1 - P_{01})/(P_{01} + P_{10}) \tag{4.26}$$

$$(101) = (P_{01})^2 P_{10}/(P_{01} + P_{10}) \tag{4.27}$$

$$(010) = P_{01}(P_{10})^2/(P_{01} + P_{10}) \tag{4.28}$$

$$(011) + (110) = 2P_{01}P_{10}(1 - P_{10})/(P_{01} + P_{10}) \tag{4.29}$$

$$(111) = P_{01}(1 - P_{10})^2/(P_{01} + P_{10}) \tag{4.30}$$

Equations (4.25)–(4.30) contain just two independent variables and five dependent variables from six observations; thus this system of equations is sufficiently overdetermined to test for first-order Markovian behavior.

Equations (4.25)–(4.30) have other advantages because they can also be used to establish the necessary dyad–triad–n-ad relationships. For example, the validity of the relationships given by Eqs. (3.24)–(3.31) in Chapter 3 can be demonstrated through substitution of the probability definitions.

As discussed in Section 4.1, establishment of conformity to first-order Markov (or any higher model) leads to a complete structural determination of the polymer examined. For example, the number-average sequence lengths are given by substituting appropriate probability expressions into either Eqs. (3.14), (3.15) or (3.10), (3.11) of Chapter 3. They are

$$\bar{n}_0 = 1/P_{01} \tag{4.31}$$

and

$$\bar{n}_1 = 1/P_{10} \tag{4.32}$$

which agree with the results for first-order Markovian systems initially derived by Price (41, p. 221). When testing for conformity to first-order Markov behavior, we can use number-average sequence lengths as an alternative means for examining the internal consistency of a dyad or triad distribution. As will be seen in Section 4.3, number-average sequence lengths do provide a means for distinguishing first-order Markovian from Bernoullian fits under conditions when simple comparisons of calculated versus observed triad distributions are difficult to evaluate.

An advantage of using the first-order Markov model for fitting an observed dyad or triad distribution is that the system will reduce to the Bernoullian case when

$$P_{01} = P_{11} \tag{4.33}$$

and

$$P_{10} = P_{00} \tag{4.34}$$

thus any Bernoullian system could be described as a special case of the first-order Markov system. For distinguishing first-order Markovian from Bernoullian behavior, we have several tools at our disposal: a calculated versus observed n-ad comonomer distribution for both cases (or calculated versus observed number-average sequence lengths), an inspection of the equality of P_{01} and P_{11} and P_{10} versus P_{00}, and finally, an inspection of the triad ratios, 001/010 and 110/101, which should equal 2.0 for Bernoullian behavior.

As has been customary throughout this book, we shall depart from the mathematical development at this point and consider an application in polymer analyses. In Chapter 3 we used hydrogenated polybutadienes as copolymer examples for determining triad distributions and number-average sequence lengths. With this information, we are also in a position to examine for conformity to either Markovian or Bernoullian statistical behavior.

4.3 STATISTICAL ANALYSES OF HYDROGENATED POLYBUTADIENES

In Chapter 3, a procedure was given for determining a comonomeric triad distribution for hydrogenated polybutadienes from ^{13}C NMR intensity data. In this section, we shall be concerned with the use of an intensity distribution (or the triad distribution) to establish if the observed distribution corresponds to a Bernoullian system or a first-order Markov statistical model. As discussed in the previous section, first-order Markov fits indicate that the comonomer distributions are nonrandom.

The triad distributions for three hydrogenated polybutadienes as obtained from ^{13}C NMR have been shown in Table 3-2. These polybutadienes were prepared under carefully controlled reaction conditions to produce 72% 1,2-butadiene additions (sample A), 50% 1,2-butadiene additions (sample B), and 26% 1,2-butadiene additions (sample C), as measured by infrared analysis (33, 34). These ranges of 1,2 versus 1,4-butadiene and the fact that these polymers have been carefully prepared and well characterized establish these samples as good tests for evaluation of statistical behavior.

Initially, an examination of a triad distribution should be directed toward a calculated comonomer distribution. The sum of the 1-centered triads, 111 + 110 + 010, gives the overall percent 1,2-butadiene in the polymer. For A, we obtain 73%, for B, 47%, and for C, 25% in agreement with the reported monomer contents by infrared. Secondly, the ratios 110/101 and 001/010 tabulated in Table 4-1, should be inspected for conformity with Bernoullian behavior. As discussed in Section 4.2, a ratio of 2.0 is expected uniformly for both 110/101 and 001/010 if Bernoullian behavior is exhibited. Three ratios in Table 4-1 deviate significantly from 2.0; however, two of these three ratios were calculated from triads with relatively low concentrations which leaves in doubt the accuracy of the ratio determination. The largest experimental error is anticipated for the lowest triad concentrations,

TABLE 4-1 Triad Ratios 110/101 and 001/010 for Hydrogenated Polybutadienes[a]

Sample	110/101	001/010
A	1.9	1.4
B	1.4	1.9
C	1.0	2.0

[a] See Randall (33).

which in turn lead to the least certain ratios. The 110/101 ratio for sample B, however, clearly departs from the expected 2.0 ratio. In spite of this observation, other criteria are needed if we wish to establish without equivocation whether any of these samples conform to Bernoullian behavior.

A comparison of calculated triad distributions initially assuming Bernoullian and then first-order Markov behavior is helpful in establishing the differences expected in the comonomer distributions. Calculated triad distributions, obtained from best fits, are given in Table 14-2 for both Bernoullian and first-order Markov models. (A second Bernoullian distribution, based on mole fractions determined directly from the triad data, is included for each polymer.) Observed triad distributions for samples A, B, and C listed in Table 4-2 appear by inspection to show better agreement with the first-order Markov statistical models. (Note that both models give satisfactory fits for sample A.) The most striking departure from Bernoullian behavior occurs for sample B which has 50% 1,2 additions from infrared analysis. The bias in the observed triad distribution from a perfect 1:2:1:1:2:1 indicates that the 1,2 additions are slightly below 50%; however, the best Bernoullian P_1 of 0.47 failed to produce as satisfactory a fit as did the best first-order Markov fit with a predicted 1,2 content of 48%. A comparison of triad distributions thus favors first-order Markov statistical behavior.

Differences observed for the conditional probabilities P_{00}, P_{10}, and P_{11}, P_{01} obtained from the best first-order Markov fits provide another method for distinguishing Bernoullian from first-order Markov behavior. A comparison of these conditional probabilities is given in Table 4-3. The first-order Markov conditional probabilities in Table 4-3 show differences that were required to predict the observed triad comonomer distribution; thus, supporting evidence is obtained for first-order Markov statistical behavior by these polybutadienes.

Finally, the calculated versus observed comonomer distributions can also be evaluated in the form of number-average sequence lengths. These results

TABLE 4-2 Calculated and Observed Triad Comonomer Distributions for Hydrogenated Polybutadienes[a]

Triad	Observed distribution	Bernoullian		First-order Markov
Sample A				
		$P_1 = 0.73$[b]	$P_1 = 0.71$[c]	$P_{01} = 0.80$ $P_{10} = 0.30$
(111)	0.36	0.389	0.358	0.36
(110)	0.30	0.288	0.292	0.31
(010)	0.07	0.053	0.060	0.07
(101)	0.16	0.144	0.146	0.17
(001)	0.10	0.106	0.119	0.09
(000)	0.01	0.020	0.024	0.01
Sample B				
		$P_1 = 0.50$	$P_1 = 0.47$[b,c]	$P_{01} = 0.54$ $P_{10} = 0.58$
(111)	0.09	0.125	0.104	0.09
(110)	0.23	0.250	0.234	0.23
(010)	0.15	0.125	0.132	0.16
(101)	0.16	0.125	0.117	0.15
(001)	0.28	0.250	0.264	0.26
(000)	0.09	0.125	0.149	0.11
Sample C				
		$P_1 = 0.25$[b]	$P_1 = 0.29$[c]	$P_{01} = 0.29$ $P_{10} = 0.81$
(111)	0.02	0.016	0.024	0.01
(110)	0.07	0.094	0.119	0.08
(010)	0.16	0.141	0.146	0.17
(101)	0.06	0.047	0.060	0.06
(001)	0.32	0.281	0.292	0.30
(000)	0.36	0.422	0.358	0.37

(Bernoullian and First-order Markov columns are under the heading "Calculated distribution".)

[a] See Randall (33).
[b] Mole fraction of 1,2-butadiene additions calculated from the triad distribution.
[c] By iterating over P_0 (best fit).

appear to be conclusively first-order Markov as shown in Table 4-4. As a final check, triad ratios calculated from the first-order Markov fits are compared to the observed ratios discussed initially. The results shown in Table 4-5 again are consistent with a first-order Markov fit. Number-average sequence lengths and triad ratios appear to offer better conformity tests for

TABLE 4-3 First-Order Markov Conditional Probabilities Based on Best Fit for Samples[a]

Sample	P_{00}	P_{10}	P_{11}	P_{01}
A	0.20	0.30	0.70	0.80
B	0.46	0.58	0.42	0.54
C	0.71	0.81	0.19	0.29

[a] See Randall (33).

TABLE 4-4 Observed versus Calculated Number-Average Sequence Lengths for 1,4 and 1,2 Butadiene Additions in Hydrogenated Polybutadienes[a]

		Calculated	
Sample	Observed[b]	First-order Markov	Bernoullian
	Number-average sequence lengths of 1,2 additions		
A	3.32 ± 0.10	3.33	3.45
B	1.78 ± 0.05	1.72	1.89
C	1.28 ± 0.02	1.23	1.41
	Number-average sequence lengths of 1,4 additions		
A	1.32 ± 0.02	1.25	1.41
B	1.78 ± 0.02	1.85	2.13
C	3.20 ± 0.12	3.45	3.45

[a] Assuming Markovian and Bernoullian behavior (33).
[b] ± Standard deviation shown for observed values.

TABLE 4-5 Comonomer Triad Ratios 110/101 and 001/010 for Polybutadiene Samples[a]

	110/101		001/010	
Sample	Observed	Calculated	Observed	Calculated
A	1.9	1.8	1.4	1.3
B	1.4	1.5	1.9	1.6
C	1.0	1.3	1.8	2.0

[a]Based on first-order Markov statistical behavior (33).

either first-order Markov or Bernoullian behavior than do direct comparisons of the observed and calculated comonomer triad distributions.

Taken collectively, the results indicate that the better fit is the first-order Markov statistical model. This result is important because it suggests that during polymerization, monomer additions are affected by the outcome of the immediately preceding addition. A second conclusion from this statistical study is that an opposite sequence of additions is favored over a like sequence of additions, that is, $P_{01} > P_{11}$ and $P_{10} > P_{00}$. Although the power of Markov statistical fits is illustrated by this example, such analyses should be approached cautiously because we have also demonstrated that small differences may exist between first-order Markov and Bernoullian data fits.

4.4 STATISTICAL ANALYSES OF CONFIGURATIONAL SEQUENCES IN VINYL HOMOPOLYMERS

Historically, the Markovian and Bernoullian statistical models for polymers were developed for applications in tacticity studies of vinyl polymers (48, 49). Price and Bovey pioneered the statistical development through equivalent statistical approaches. Mathematically, the Price development differs from that of Bovey since the states used in the statistical analysis are not the same. We shall explore both approaches in detail beginning with that of Price because it is analogous to the development for copolymers in the preceding sections.

In Section 4.1, the Bernoullian model was developed for an application to copolymers with units defined by 0 and 1. An analogous situation exists for the two types of configurational additions in vinyl polymers. A 0 can be used to identify one mode of addition while a 1 identifies the opposite, because from a configurational viewpoint, a vinyl homopolymer can be considered a copolymer. There is a significant departure, however, from the copolymer mathematical treatment as applied to vinyl homopolymers. Only like and opposite additions can be distinguished; thus the 00 addition is the same as the 11 addition. Price, in his treatment, used different probabilities for 1 and 0 additions and therefore could define the mathematical framework in terms of one-unit statistical states. After a complete development, equalities among various conditional probabilities were duly noted because of stereochemical equivalence that reduced the results to a simpler mathematical form.

Let us examine the statistical framework of Price using one-unit states for both Bernoullian and first-order Markov statistical treatment as shown in the accompanying tabulation.

Initial state	Add	Final state	Bernoullian conditional probability	
0	0	0	P_0	
1	0	0	P_0	
				$P_0 + P_1 = 1$
0	1	1	P_1	
1	1	1	P_1	
			First-order Markov conditional probability	
0	0	0	P_{00}	$P_{00} + P_{01} = 1$
0	1	1	P_{01}	
1	0	0	P_{10}	$P_{11} + P_{10} = 1$
1	1	1	P_{11}	

As defined in the copolymer discussion in Section 4.1, the Bernoullian conditional probability and the probability of finding a specific unit are the same because the conditional probability is independent of an outcome of a previous event. The relative dyad concentrations in terms of Bernoullian conditional probabilities are

$$(00) = P_0{}^2 \tag{4.35}$$

$$(01) = P_0 P_1 = P_0(1 - P_0) \tag{4.36}$$

$$(10) = P_1 P_0 = P_0(1 - P_0) \tag{4.37}$$

$$(11) = P_1{}^2 = (1 - P_0)^2 \tag{4.38}$$

However, only two configurational dyads are possible, a like combination and an unlike combination; therefore, a 00 sequence is equivalent to a 11 sequence and 01 is equivalent to 10. Under these circumstances, there is a unique solution to Eqs. (4.35)–(4.38), that is,

$$P_0 = P_1 = 1/2 \tag{4.39}$$

and Price's Bernoullian description fits any polymer that is "ideally random (49)." There are equal numbers of monomer units with opposite configurations and the sequence concentrations are appropriate multiples of $\frac{1}{2}$.

For a first-order Markov system, the relative dyad concentrations are

$$(00) = {}^1P_0P_{00} \tag{4.40}$$

$$(01) = {}^1P_0P_{01} \tag{4.41}$$

$$(10) = {}^1P_1P_{10} \tag{4.42}$$

$$(11) = {}^1P_1P_{11} \tag{4.43}$$

where ${}^1P_0$ and ${}^1P_1$ represent the probability of finding 0 and 1 placements, respectively. As was the case for A–B type copolymers,

$${}^1P_0 + {}^1P_1 = 1 \tag{4.44}$$

and

$${}^1P_0P_{01} + {}^1P_1P_{11} = {}^1P_1 \tag{4.45}$$

$${}^1P_1P_{10} + {}^1P_0P_{00} = {}^1P_0 \tag{4.46}$$

which leads to

$${}^1P_1 = P_{01}/(P_{01} + P_{10}) \tag{4.47}$$

$${}^1P_0 = P_{10}/(P_{01} + P_{10}) \tag{4.48}$$

However, as in the Bernoullian analysis, the equivalence of like and opposite placements must be noted, that is,

$$P_{00} = P_{11} \quad \text{and} \quad P_{01} = P_{10} \qquad (4.49), (4.50)$$

and Eqs. (4.47) and (4.48) become

$${}^1P_0 = {}^1P_1 = \tfrac{1}{2} \tag{4.51}$$

For the relative dyad concentrations, the first-order Markov definitions are

$$(11) + (00) = 2\,{}^1P_0P_{00} = P_{00} \tag{4.52}$$

$$(10) + (01) = 2\,{}^1P_0P_{01} = 1 - P_{00} \tag{4.53}$$

Corresponding treatments lead to the triad equations,

$$(111) + (000) = 2\,{}^1P_0P_{00}P_{00} = P_{00}^2 \tag{4.54}$$

$$(110) + (011) + (100) + (001) = 4\,{}^1P_0P_{00}P_{01} = 2P_{00}(1 - P_{00}) \tag{4.55}$$

$$(010) + (101) = 2\,{}^1P_1P_{10}P_{01} = (1 - P_{00})^2 \tag{4.56}$$

and tetrad equations,

$$(1111) + (0000) = 2^1 P_0 P_{00} P_{00} P_{00} = P_{00}^3 \quad (4.57)$$

$$(1110) + (0111) + (1000) + (0001) = 4^1 P_0 P_{00} P_{00} P_{01}$$
$$= 2P_{00}^2(1 - P_{00}) \quad (4.58)$$

$$(0110) + (1001) = 2^1 P_1 P_{10} P_{00} P_{01}$$
$$= P_{00}(1 - P_{00})^2 \quad (4.59)$$

$$(1101) + (1011) + (0100) + (0010) = 4^1 P_0 P_{00} P_{01} P_{10}$$
$$= 2P_{00}(1 - P_{00})^2 \quad (4.60)$$

$$(1100) + (0011) = 2^1 P_0 P_{00} P_{01} P_{11}$$
$$= P_{00}^2(1 - P_{00}) \quad (4.61)$$

$$(0101) + (1010) = 2^1 P_1 P_{10} P_{01} P_{10}$$
$$= (1 - P_{00})^3 \quad (4.62)$$

In the Price development, we can see that Bernoullian systems are defined by a unique conditional probability of $\frac{1}{2}$ while the first-order model is defined by only one independent variable P_{00}.

In contrast to the one-unit state probability models presented above, Bovey used two-unit states to define the statistics of polymer configurations. A vinyl polymer is defined as a copolymer consisting of *meso* and *racemic* dyads and the Bernoullian probability for a *meso* addition is simply P_m and a *racemic* addition, $1 - P_m$. Stereochemical equivalence is built into the system and no equivalent mathematical states develop during formulation. Equations for dyad through tetrad distributions can be derived more easily than with the Price one-unit state.

For dyads:

$$(m) = P_m \quad (4.63)$$

$$(r) = 1 - P_m \quad (4.64)$$

for triads:

$$(mm) = P_m{}^2 \quad (4.65)$$

$$(rm) + (mr) = 2(1 - P_m)P_m \quad (4.66)$$

$$(rr) = (1 - P_m)^2 \quad (4.67)$$

and for tetrads:

$$(mmm) = P_m{}^3 \tag{4.68}$$

$$(rmm) + (mmr) = 2P_m{}^2(1 - P_m) \tag{4.69}$$

$$(rmr) = P_m(1 - P_m)^2 \tag{4.70}$$

$$(mrm) = P_m{}^2(1 - P_m) \tag{4.71}$$

$$(rrm) + (mrr) = 2P_m(1 - P_m)^2 \tag{4.72}$$

$$(rrr) = (1 - P_m)^3 \tag{4.73}$$

Note that Eqs. (4.63)–(4.73) are identical in form to Eqs. (4.52)–(4.62) derived previously for the one-unit states using the Price framework. However, Eqs. (4.63)–(4.73) were developed from a Bernoullian model, whereas Eqs. (4.52)–(4.62) were derived for first-order Markov. These results do not suggest that one of these approaches is incorrect but do show how the state definitions affect the statistical development. The order in statistical treatments is defined with respect to the states used in the mathematical development and, therefore, does not describe some particular polymer behavior. The same physical interpretation results from both the Price and Bovey treatments. In the statistical analyses presented by Bovey, *meso* and *racemic* dyads form the states used in the statistical development. For systems conforming to Bernoullian behavior, the probability of a *meso* or *racemic* addition is, therefore, independent of the configuration of the previous additions. The first-order Markov development of Price describes a system that depends upon the outcome of the immediate addition but does not depend upon the previous configuration. Two successive polymer units are required to define configurational relationships. With the Price one-unit states, steric effects associated with monomer configurations occur for the second-order Markov model. In either case, a configurational dependence occurs when an immediate pair of additions affect a subsequent addition irrespective of the order defined in the statistical analysis.

The Bernoullian definition evolving from the Price mathematical framework can become confusing because it is not the same Bernoullian definition used by Bovey. With the Price one-unit states, the only Bernoullian system occurs when $P_{00} = \frac{1}{2}$, which corresponds to the ideally random Bernoullian system of Bovey where $P_m = \frac{1}{2}$. Systems where $P_m \neq \frac{1}{2}$ have been called biased Bernoullian (50) and correspond to the Price first-order Markov development. Thus in the Price treatment, the Bernoullian system corresponds to the ideally random system in the two-unit state development and first-order Markov is the same as biased Bernoullian in the Bovey treatment.

As far as the polymer chemist is concerned, the concept of *order* is immaterial because any interpretation of the statistical behavior must be made with an awareness of the states used in the mathematical framework. The concept of *order*, therefore, does not infer any particular reaction scheme for the polymer (41, p. 237). We should also be reminded that this particular development arose because of an equivalence among configurational states for vinyl homopolymers and is not a factor in Bernoullian and Markovian analyses of comonomer arrangements in copolymers and terpolymers.

Finally, the mathematical probability scheme for sterochemical descriptions in vinyl homopolymers followed by most NMR spectroscopists is that of Bovey. Therefore, the two-unit state model for both Bernoullian and Markovian treatments suggested by Bovey will be used throughout the remainder of our discussion on polymer tacticity. The comparison of the Price and Bovey treatments underscores the need for more than a cursory look at the mathematical framework used in probability models.

4.5 A STATISTICAL ANALYSIS OF CONFIGURATIONAL SEQUENCES IN ATACTIC POLYSTYRENES

Statistical analyses have been used to make ^{13}C chemical shift assignments in poly(vinyl chloride) (46), poly(vinyl alcohol) (47), poly(vinyl acetate) (51) and polystyrene (19). The method offers one of the more interesting assignment techniques; however, results are not always unambiguous. Conflicting conclusions, which depended upon the spectral region and the detail of the analysis, have been reported for polystyrene.† The methylene resonances of polystyrene offer an interesting example. The ^{13}C NMR spectrum of the methylene and methine region of a free radical polymerized polystyrene is shown in Fig. 4.1. The methylene resonances are numbered from 1 to 9 while resonance 10 is the only methine resonance (52). The relative areas of resonances 1–9 are given in Table 4-6 as obtained by cutting and weighing each peak and by curve fitting using Lorentzian peak shapes.

An analysis of the methylene resonance assignments in polystyrene begins with a determination of the chemical shift sensitivity. Nine resonances are observed when a pure tetrad chemical shift sensitivity would produce a maximum of six resonances and a pure hexad chemical shift sensitivity would

† See Section 6.6 for a discussion of the assignments in polystyrene.

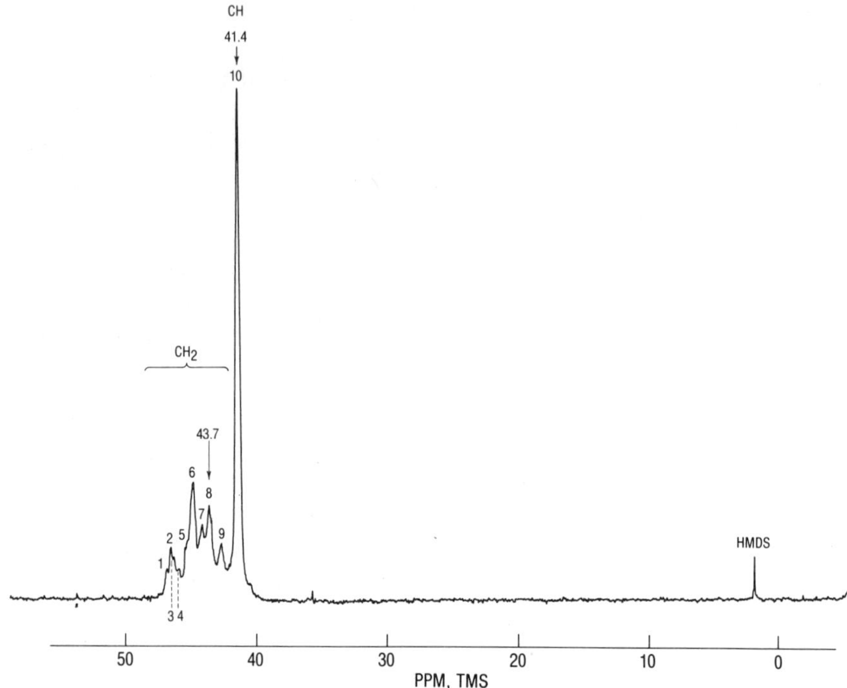

Fig. 4.1. Proton noise decoupled ^{13}C NMR spectrum at 25.2 MHz of a free radical initiated polystyrene in 1,2,4-trichlorobenzene at 120°C.

TABLE 4-6 Relative Areas of Polystyrene Methylene Resonances[a]

	Relative area	
Peak	By cutting and weighing	By curve fitting
1	0.05	0.04
2	0.07	0.09
3	0.04	0.04
4	0.05	0.03
5	0.10	0.11
6	0.28	0.31
7	0.13	0.12
8	0.19	0.19
9	0.09	0.07

[a] Determined by using pure Lorentzian peak shapes (19, 27).

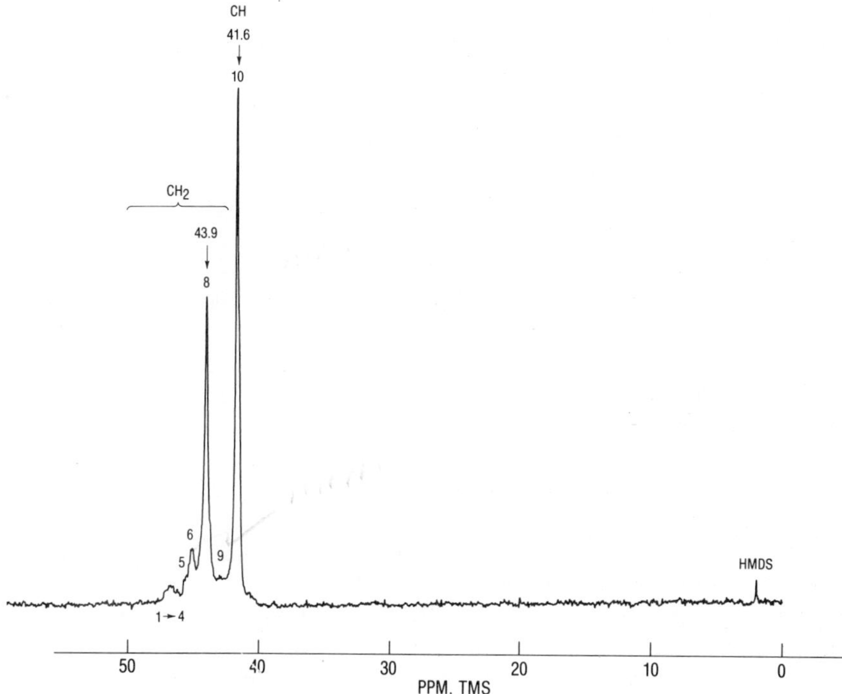

Fig. 4.2. Proton noise decoupled ^{13}C NMR spectrum at 25.2 MHz of a polystyrene containing predominantly isotactic sequences in 1,2,4-trichlorobenzene at 120°C.

produce 20 resonances. We are faced, therefore, with a problem of either unraveling a mixture of closely spaced tetrad and hexad resonances or resolving a set of overlapping hexad resonances.

An assignment for isotactic sequences can be made reasonably; however, the length of the sequence detected must be established. A spectrum of a predominantly isotactic polystyrene is shown in Fig. 4.2. Resonance 8 can be identified as arising from methylene carbons in isotactic sequences; however, we do not know whether resonance 8 originates from dyad, tetrad, or even hexad isotactic sequences. One approach that leads to an isotactic sequence length assignment utilizes the initial assumption that these polystyrene sequence distributions are Bernoullian (53). Subsequently, the relative area of resonance 8 can be used to calculate a P_m for either a dyad, tetrad, or hexad sequence, as shown:

Isotactic sequence length	P_m
Dyad (*m*)	0.19^1 or 0.19
Tetrad (*mmm*)	$(0.19)^{1/3}$ or 0.575
Hexad (*mmmmm*)	$(0.19)^{1/5}$ or 0.717

Relative concentrations of the remaining configurational sequences can then be calculated for any particular combination of sequence lengths.

Table 4-7 contains the remaining tetrad concentrations that were calculated from each of the P_m values. A total of only four peaks are predicted if the isotactic peak represents a dyad and the remaining *r*-centered sequences are tetrad sensitive. Correspondingly, a total of eight peaks are predicted if the isotactic sequence is hexad sensitive and the remaining sequences are tetrad sensitive. Six peaks, of course, are anticipated for a pure tetrad sensitivity. Each of these situations is included in Table 4-7. A comparison of the observed relative methylene peak intensities with the various distributions calculated in Table 4-7 shows a reasonable fit for the pure tetrad distribution if peaks 1–4 are combined to give a single tetrad. These results are given in Table 4-8. The fit is strengthened by the observation that four peaks were combined to give the *rrm* relative tetrad concentration. The *rrm* sequence is only one of two tetrads that gives four hexad sequences as shown in Fig. 1.7. The calculated *rrm*-centered relative hexad concentrations and observed relative intensities for peaks 1–4 are given in Table 4-9. Overall, a satisfactory fit is obtained assuming that the methylene resonances result from the four *rrm*-centered hexads plus the five remaining tetrads. This result, however, can be criticized from the viewpoint that free radical polymerized polystyrenes have predominantly syndiotactic structures (54, 55). The 0.575 value of P_m, which gave the satisfactory methylene fit, indicates slightly more isotactic than syndiotactic sequences.

In a corresponding analysis of the substituted aromatic carbon resonances, Matsuzaki *et al.* (56) obtained a fit which indicated that free radical and *n*-butyl lithium polymerized polystyrenes were predominantly syndiotactic. A detailed analysis, however, was not performed because the 20 to 21 carbon resonances of the aromatic region (see Fig. 6.7) were divided into three regions that allowed a fit with only two independent experimental observations. In spite of this possible oversimplification, a P_m of 0.20, which is in agreement with that calculated assuming a methylene dyad chemical shift sensitivity, was obtained for free radical polymerized polystyrenes. It is possible to satisfy the requirement of nine methylene resonances, if in addition to the *m* dyad, there are four *rrm*-centered hexads, three *rrr*-centered hexads, and an *mrm* tetrad. The intensity distribution calculated with a P_m of 0.20, however, does not fit the observed intensity distribution as well as that calculated with a P_m of 0.575.

The assignments given for the methylene resonances of the free radical polymerized polystyrene (Tables 4-8 and 4-9) should be considered tentative until confirmation is obtained. Because the chances for a coincidental fit are so great, assignments based on Bernoullian fits should be supported by

TABLE 4-7 Predicted Tetrad Relative Concentrations As a Function of Isotactic Chemical Shift Sensitivity

Sensitivity					
Isotactic dyad (P_m = 0.190)		Isotactic tetrad (P_m = 0.575)		Isotactic hexad (P_m = 0.717)	
m	0.19	*mmm*	0.19	*mmmmr*	0.15
				mmmmm	0.19
				rmmmr	0.03
		mmr	0.28	*mmr*	0.29
		rmr	0.06	*rmr*	0.06
rrr	0.53	*rrr*	0.08	*rrr*	0.02
rrm	0.25	*rrm*	0.21	*rrm*	0.11
mrm	0.03	*mrm*	0.14	*mrm*	0.15

TABLE 4-8 Calculated and Observed Polystyrene Methylene ^{13}C NMR Instensities[a]

	Relative area			
Peak	Observed		Calculated (P_m = 0.575)	
1–4	0.21	0.20	0.21	(*rrm*)
5	0.10	0.11	0.10	(*rmr*)
6	0.28	0.31	0.28	(*mmr*)
7	0.13	0.12	0.14	(*mrm*)
8	0.19	0.19	0.19	(*mmm*)
9	0.09	0.07	0.08	(*rrr*)

[a] Assuming a basic tetrad chemical shift sensitivity (19).

TABLE 4-9 Calculated *rrm*-Centered Hexad Sequence Concentrations and Relative Areas of Polystyrene Methylene Resonances 1–4[a]

	Relative area			
Peak	Observed		Calculated (P_m = 0.575)	
1	0.05	0.04	0.05	(*mmrrr*)
2	0.07	0.09	0.07	(*mmrrm*)
3	0.04	0.04	0.04	(*rmrrr*)
4	0.05	0.03	0.05	(*rmrrm*)

[a] See Randall (19).

either model compounds, model polymers, or conformational data and arguments (46).

In summary, Markov statistics offer one of the more interesting approaches for making NMR chemical shift assignments and for determining concentrations of specific comonomer sequences. Because both the analyses and results are often intriguing, one must eastablish if a fit has described a real polymer system and not some artifact created by either coincidence or too few experimental observations for a meaningful analysis. The number of degrees of freedom in an analysis is always a consideration because a Markov model of sufficiently high order will eventually fit almost any imaginable polymer system.

Another problem facing the polymer chemist performing Markov statistical analyses is whether the polymerization was homogeneous throughout; that is, whether polymer was produced under the same reaction conditions during the initial steps of polymerization as when the polymerization was completed. Any subsequent Markovian analysis will be based on an average structure of polymer produced under different reaction conditions. It has also been proposed that different polymers can be produced concurrently (23, 57) at different catalyst sites; thus any statistical fit would be based on an average of different structures and the conditional probabilities obtained would not reflect the true experimental conditions. Free radical polymerizations and polymerizations performed under carefully controlled reaction conditions offer the best opportunities for meaningful statistical analyses. Thus Markov statistics must be handled carefully. Because complete structural characterizations are obtained, it is easy to become overwhelmed by the magnitude of the results.

5 Experimental Design for Quantitative FT-NMR Measurements

In Chapters 1–4, a ^{13}C NMR polymer spectrum was the beginning point of the discussion. It was assumed that NMR data were obtained in such a way to portray an accurate picture of polymer structure. Assignments were made and the origins of signals were defined in terms of contributing structures. Finally, it was shown how monomer distributions and number-average sequence lengths could be obtained from NMR intensity data.

In this chapter, we shall consider the NMR experiment from a standpoint of data accumulation and intensity measurements. It will be assumed that the reader is familiar with the basic Fourier transform concepts although the material will be presented in a manner that allows the reader to grasp the fundamentals without being an expert in NMR. Emphasis will be placed upon the experimental requirements and design for reliable quantititative measurements rather than the mathematical formulations involved in NMR Fourier transform (FT-NMR) analyses.

5.1 SATURATION AND RESOLUTION IN FT-NMR

A Lorentzian in the NMR frequency domain and an exponential in the NMR time domain form a Fourier transform pair (2, p. 15). In conventional continuous wave (cw) NMR, the Lorentzian signals are observed directly by selective excitement of nuclei by either magnetic field sweep or rf frequency sweep. In quantitative studies, care is given to avoid saturation of NMR signals and to achieve maximum resolution through good field homogeneity. These same standards apply to NMR exponential gathered in the time domain and converted to the frequency domain via the Fourier transform technique; however, the safeguards in FT-NMR are handled in a completely different manner.

For naturally abundant ^{13}C NMR spectra, FT-NMR is a highly efficient method. The NMR experiment in the time domain involves a collective excitement of the ^{13}C nuclei from a previously established equilibrium position in a magnetic field. The number of nuclei perturbed from equilibrium is determined by the strength or duration of an rf pulse. Once the rf pulse is removed, the excited ^{13}C nuclei return to the former equilibrium position by a thermal relaxation process described as free induction decay (FID). It is the FID that is observed in the time domain and converted to the frequency domain by the Fourier transform technique. The appearance and the duration of FID exponentials depends upon two factors: (a) the difference in frequency between the applied rf field and the precessional frequencies (Lamor) of the ^{13}C nuclei rotating about the field axis, and (b) the rate at which the nuclei return to equilibrium. A pure exponential is obtained for an FID if it is pulsed precisely at its Larmor frequency. Normally with off-resonance excitation the FID is a composite of signals that appears to ring as observed in conventional high resolution continuous wave NMR experiments. The appearance of the resultant FID exponential and its complexity depend upon the number of different types of nuclei excited and the relative difference between each Larmor frequency and the applied rf field. The rate of relaxation is given by the inverse of the spin–lattice relaxation time which is the time constant for decay of magnetization components along the field direction. A single pass spectrum, which would take several minutes by cw NMR, can be obtained in a fraction of a second or in only a few seconds. Generally, a train of rf pulses is applied during a ^{13}C NMR experiment and the FID's are averaged to improve the signal-to-noise ratio.

We may now inquire how resolution and saturation can be controlled in an FT-NMR experiment. If a train of rf pulses is so closely spaced that full relaxation does not occur between pulses, a result will be obtained that is analogous to saturation in the cw mode. Maximum intensities will not be realized; instead, the signal intensities will be proportional to the extent of relaxation that may not be the same throughout the spectrum. For a 90° pulse (or 90° tip angle of the nuclear magnetization components away from equilibrium), the pulse spacings must be five times the spin–lattice relaxation time (T_1) to ensure 99% relaxation (2, p. 21). In the collection of FID's for quantitative experiments, the pulse spacings must be five times the T_1 of the slowest relaxing nucleus if true relative intensities are to be realized. The pulse spacings can be controlled by inserting time delays between pulses. In Fig. 5.1, polypropylene spectra, which were obtained with pulse spacings of 1 and 10 sec, are shown to illustrate the importance of pulse spacings. It is immediately obvious from the effects of these pulse spacings upon the observed intensities that the methyl carbon has the longest T_1 and the methylene carbon has the shortest. These spectra were recorded at 120°C in 1,2,4-trichlorobenzene where the methyl carbon has a T_1 of 2.1 sec, the methine carbon 1.5 sec, and the methylene carbon 0.7 sec (58, 59). If the relaxation times were unknown, the proper pulse spacings would necessarily have to be determined by trial and error. A poor choice of pulse spacings also leads to a less efficient experiment because the methylene carbon

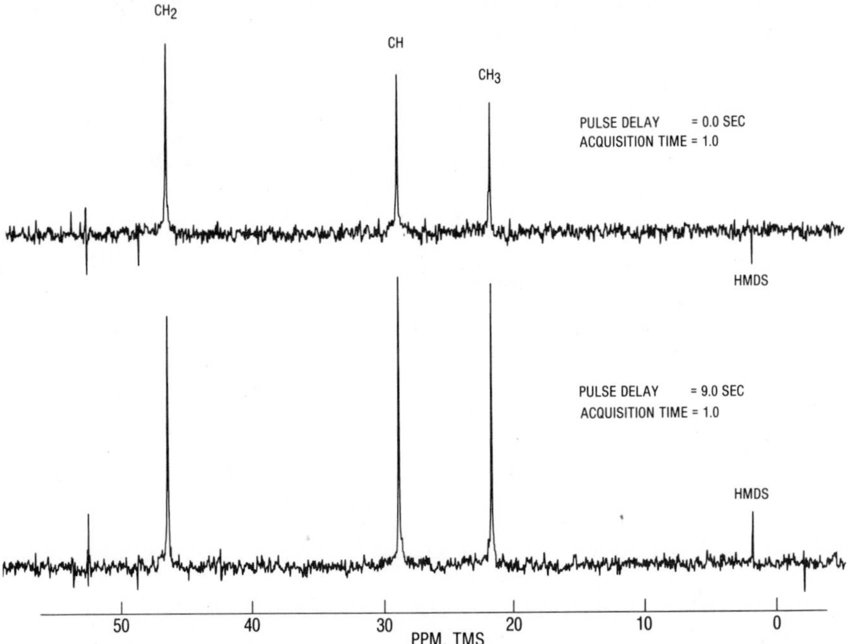

Fig. 5.1. Proton noise decoupled ^{13}C NMR spectra at 25.2 MHz of the polypropylene of Fig. 1.3 obtained with pulse spacings of 1.0 and 10.0 sec.

resonance, which is the strongest of the three resonances in the experiment with the 1.0 sec pulse spacing, still shows a 24% intensity loss when compared to the experiment with 10.0 sec between pulses.

A second experimental factor of concern in FT-NMR experiments for quantitative applications is resolution. It is possible in FT-NMR experiments to mask good resolution from a well-tuned magnetic field by improperly recording FID exponentials. It is usually unnecessary, if not undesirable, to record the entire FID during data accumulation. The computer is allowed to record only the initial free induction decay for a period of time referred to as the *acquisition time* AT. The maximum resolution that can be observed in the resultant spectrum is given by the reciprocal of the acquisition time 1/AT. Spectra of an amorphous polypropylene, shown in Fig. 5.2, illustrate the differences in resolution achieved with acquisition times of 2.0 and 0.2 sec for an equal number of FID's and a total spacing of 10 sec between pulses. Clearly, a measurement of relative intensities will be made difficult, if not unreliable, through an improper choice of acquisition time. An advantage of the shorter acquisition time, apparent in Fig. 5.2, is that a better signal-to-noise ratio is achieved. Such an experimental setup would be reasonable for polymer spectra where only well-separated lines are observed. In each case, one must decide the minimum resolution requirements for quantitative measurements.

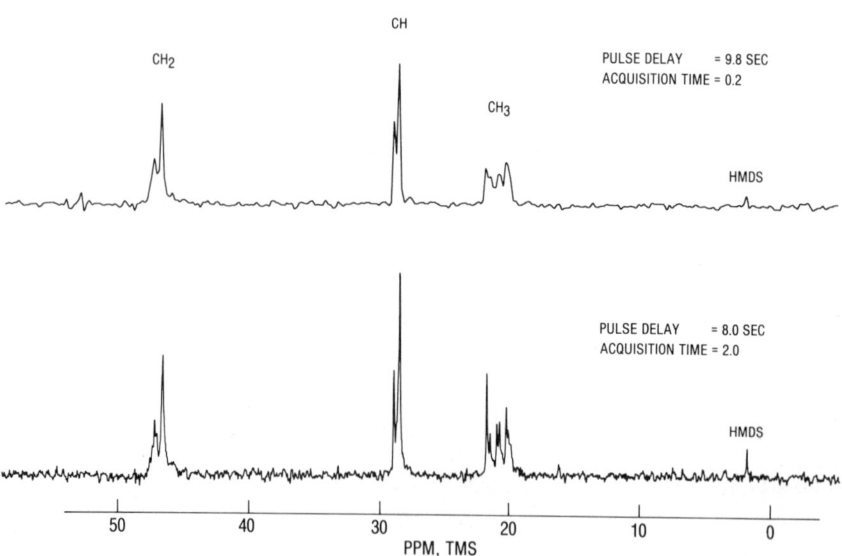

Fig. 5.2. Proton noise decoupled ^{13}C NMR spectra at 25.2 MHz of the amorphous polypropylene of Fig. 1.4 obtained with acquisition times of 0.2 and 2.0 sec.

With the proper experimental setup, that is, an acquisition time of at least 1.0 sec and a total pulse spacing of approximately $5T_1$ for the slowest relaxing nucleus, NMR data that have no induced discrepancies because of instrumental variables can be obtained for quantitative measurements.

5.2 THE NUCLEAR OVERHAUSER EFFECT AND CARBON-13 NATURAL ABUNDANCE

In the preceding section, we have shown how NMR spectra could be affected by a combination of intrinsic variables and instrumental parameters during FT-NMR experiments. There are other intrinsic properties to be considered also; one is the nuclear Overhauser effect (NOE), and another is the natural abundance of ^{13}C nuclei.

As discussed in Chapter 1, ^{13}C NMR spectra are usually obtained with the ^{1}H–^{13}C spin–spin coupling completely removed by broad band or noise decoupling at the proton resonance frequency. Under these conditions, energy transfer can occur among the ^{1}H and ^{13}C nuclear spin levels. For these particular nuclei, an enhancement called the nuclear Overhauser effect is observed for the intensities of the ^{13}C resonances. In small molecules with unrestricted segmental mobilities, the NOE varies from factors of one to three according to the structural environments. Nuclear Overhauser effects, therefore, must be examined in quantitative studies because differences must be accounted in an analytical procedure. In polymers, Schaefer and Natusch (30) have shown that NOE's are generally the same throughout because of restricted segmental mobilities; even so, measurements of the NOE are usually desirable. NOE measurements can be accomplished through either gated decoupling, that is, the decoupler is off during the pulse and on during acquisition so that decoupled spectra are obtained without an NOE, or for some polymers, by the addition of a paramagnetic material to quench the NOE (60). The latter procedure has an additional advantage that relaxation times can be shortened with quenching that permits a faster, more efficient FT-NMR experiment. Iron(III) acetylacetonate and iron(III) trifluoroacetylacetonate are effective reagents for NOE quenching in polymer spectra (22, 61). In either case, both the NOE and T_1 should be measured under the conditions used for the quantitative NMR measurements.

An intrinsic property seldom considered in quantitative FT-NMR ^{13}C experiments is the relative natural abundance (approximately 1.1%) of ^{13}C nuclei. In quantitative applications, the ^{13}C natural abundance is usually considered to be the same throughout a given molecule. Kinetic isotope studies (62, 63) in research on petroleum genesis have suggested that the end carbons of naturally occurring hydrocarbons have slightly fewer (91 to 97% of natural abundance) ^{13}C nuclei than do carbons in interior sequences. This difference could be a factor in quantitative NMR studies of polymers if a natural product were used in the synthesis.

The three resonances in the polypropylene spectrum in Fig. 5.1 do not have exactly the same relative intensities. The methine carbon shows the strongest intensity of the three resonances and the methylene resonance shows the lowest relative intensity. This spectrum was obtained with a 90° pulse and a 10 sec pulse delay that is adequate for the observed relaxation times. Longer pulse delays do not affect this observed difference.† Of course, there are experimental variables to consider that affect the readout of relative intensities; however we should be aware of possible intrinsic differences among natural abundances. Generally, in ^{13}C NMR spectra there is considerable duplication of information as we have seen in several of the examples given earlier; thus it would be reasonable to select those resonances where ambiguities, either from differences in NOE's, T_1's, ^{13}C natural abundance, or experimentally derived factors, can be avoided in a quantitative measurement. For example, in measurements of branch content in low density polyethylenes, it is possible to develop a method that uses only backbone carbons (64). In the previous example of hydrogenated butadiene–styrene copolymers, the styrene methine carbon could be used to obtain a triad distribution, where the styrene unit was always the center unit in the triad. In general, it may be wise to avoid resonances from methyl carbons except in those tacticity studies where an internal distribution can be obtained without equivocation.

5.3 INTENSITY MEASUREMENTS IN FT-NMR QUANTITATIVE STUDIES

Intensity measurements in ^{13}C NMR polymer spectra can be made using either relative peak heights or peak areas. Peak heights can be used reliably if there is no overlap and if the peaks measured have the same line width at one-half the maximum peak height. These criteria are not often met in ^{13}C NMR polymer spectra because of a characteristic ^{13}C sensitivity to subtle structural features; thus intensity measurements are more reliable if

† This difference, conceivably, could be related to nonuniformities in the pulse power distribution.

based on relative peak areas. Spectral integration, cutting and weighing, and curve fitting can be used to obtain relative peak areas; the best approach, however, may be through curve fitting to avoid unwanted contributions from peak overlap.

NMR peak shapes, theoretically are Lorentzian (65) and a Lorentzian shape will be observed if the peak widths are determined by the transverse or T_2 relaxation process rather than field inhomogeneity. Contributions to the line widths from magnetic field inhomogeneity usually produce Gaussian peak shapes. Linewidths in ^{13}C NMR polymer spectra are generally 2 Hz or larger. Contributions from field inhomogeneity, if the field is well tuned, will be only a few tenths of a hertz; thus polymer line shapes are predominantly Lorentzian. Curve fitting results using pure Lorentzian line shapes are shown in Fig. 5.3 for the methyl region of an amorphous polypropylene discussed in earlier chapters. Results from both peak height measurements and cutting and weighing are included with curve resolving in Table 5-1 for the polypropylene spectrum in Fig. 5.3. The curve fitting method produces the most accurate results for use in calculations of the number-average sequence lengths and configurational monomer distributions.

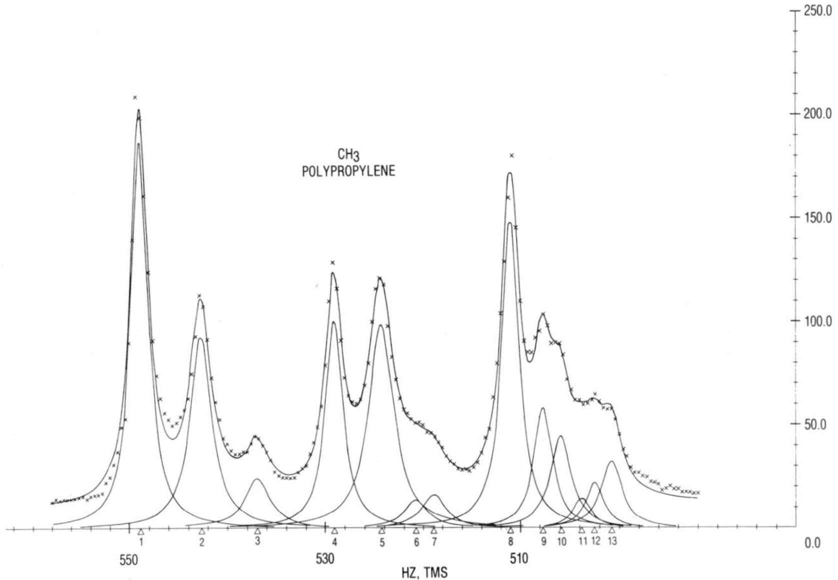

Fig. 5.3. Curve fitting results using Lorentzian peak shapes for the methyl region of the amorphous polypropylene shown previously in Fig. 1.11 (x's denote experimental points, continuous line is calculated).

TABLE 5-1 Methyl Intensity Measurements for Polypropylene[a]

Line	Relative intensities		
	Peak heights	Cutting and weighing	Curve resolving
1	0.24	0.21	0.19
2	0.10	0.12	0.12
3	0.03	0.04	0.04
4	0.12	0.11	0.11
5	0.11	0.12	0.15
6	0.04	0.04	0.02
7	0.03	0.02	0.02
8	0.17	0.16	0.16
9	0.10	0.10	0.14[b]
10	0.05	0.08	0.06[c]

[a] Shown in Fig. 5.3 (27).
[b] Sum of areas 9 and 10 in Fig. 5.3.
[c] Sum of areas 11, 12 and 13 in Fig. 5.3.

5.4 COMPUTER REQUIREMENTS IN FT QUANTITATIVE STUDIES

The computer requirements for FT-NMR studies have been well described previously (2, pp. 72–79). Computers are used to perform Fourier transformations and to collect data for signal-to-noise averaging. We shall consider computer requirements from a viewpoint of quantitative applications, because the available dynamic range can become a critical factor in determining accuracy in ^{13}C NMR quantitative studies. The number of points used during data collection and subsequent Fourier transformation is determined, in part, by the number of computer words available. Most installations have either 8K or 16K memories with 4K set aside for the FT-NMR program. Data accumulations for signal-to-noise improvements can be accomplished with either floating point or fixed point arithmetic. The former offers no problems with dynamic range but is more costly and less efficient than a fixed point system. The latter is probably used more often; however, the limitations placed upon dynamic range must be well understood for quantitative NMR studies. From a quantitative standpoint, it is the length of the word in fixed point data accumulations that is important. A typical word length is 16 bits which means that the largest integer stored as a computer word is $(2^{15} - 1)$ or 32767. (Only 15 bits are used since the 16th bit must be available for the sign.) Data from a series of repeat experiments are added coherently until a word has reached its optimum length; at that point further accumulations of data offer no improvement in the signal-to-noise ratio. This overflowing of word length can be partially avoided by using a weighted average of the form (2, p. 73).

$$\bar{x} = \bar{x} + \frac{1}{n}(x_i - \bar{x}) \qquad (5.1)$$

where $\bar{x}$ is the averaged value, x_i the newest acquired value, and n the number of samples. Another approach to avoid the overflow problem is to shift each word one bit to the right after a word has reached its maximum length which has the same effect as dividing each integer by two. The incoming information is treated in the same manner and added to the data in storage. One of the problems associated with these methods of data averaging is that NMR signals of low relative intensities will be truncated with respect to much larger signals. For example, if two NMR resonances have a relative intensity ratio of 1000:1, the first data addition will require approximately 2^{10} of the available word length. After only 31 accumulations, the word length will overflow and each word will be shifted one bit to the right. At this point, the large resonance will continue to grow, however, the small resonance will no longer be detected. A subsequent Fourier transform will produce two resonances but the relative intensities will be incorrect. This problem has been avoided in smaller 8K computer systems by collecting 10–20 free induction decay signals, performing a Fourier transform and storing the continuous wave spectrum in a previously set aside block of memory. The process is then repeated until adequate signal-to-noise is obtained through an averaging of continuous wave spectra. Truncation is avoided and true relative intensities are obtained. For those systems equipped with a disk for additional storage, double precision arithmetic can be used during data averaging which increases the available word length to $2^{31} - 1$ or 2,147,483,647. In the previous example, approximately two million FID's can be recorded without loss of the 1000:1 ratio.

One of the questions frequently asked of ^{13}C NMR spectroscopists is whether ^{13}C NMR can be used to measure branching in polyethylenes and, if so, what is the detection limit? For 8K–16K computer systems, the detection limit is near one branch per thousand carbon atoms, and only if block averaging of continuous wave spectra is used. For 16K systems equipped with disk accessories that allow double precision arithmetic during data accumulation and storage, it may be possible to detect branches in the neighborhood of one per ten thousand carbons in a reasonable time span.

The detection of branching concentrations lower than one per thousand carbons will depend upon factors other than word size. The analog-to-digital converter ADC must also have sufficient resolution, that is, 13 bits or higher, if branching in a range of one in ten thousand carbon atoms is to be detected. Thus close attention must be paid to the resolution of the ADC as well as word size for ^{13}C NMR quantitative measurements of components in less than one part in a thousand ratio.

Although the problems and pitfalls surrounding quantitative ^{13}C NMR measurements have been discussed in the preceding sections, they should not be emphasized at the expense of discouraging attempts at ^{13}C NMR quantitative studies. Carbon-13 NMR measurements are direct, with no extinction coefficients required for each type of structural entity. Measurements can be performed in such a way that each carbon type gives rise to only a single signal analogous to the molecular response in chromatography studies. Because the structural detail available from ^{13}C NMR is so extensive, an NMR spectroscopist should try to overcome those obstacles that impede quantitative measurements.

6 *A Survey of Carbon-13 NMR Studies of Vinyl Homopolymers and Copolymers*

This chapter is devoted to a brief review of published ^{13}C NMR studies of vinyl homopolymers and copolymers. A foremost consideration in most ^{13}C NMR studies of polymers is the assignment of chemical shifts that lead to determinations of comonomer sequence distributions as dyads, triads, tetrads, etc. Number-average sequence lengths have not been generally calculated from comonomer distribution data. Included in this chapter are evaluations and comparisons of chemical shift assignments in ^{13}C polymer spectra, an examination of possible trends among configurational assignments and, whenever possible, calculations of number-average sequence lengths in vinyl homopolymers and copolymers.

The approaches used in making chemical shift assignments, presented in the literature, generally depend upon the individual polymer and available supporting evidence. We may look for any internal consistencies among information derived from different spectral regions, utilize independent structural information from ^{1}H NMR, and examine spectra from reference polymers and model compounds.

Carbon-13 assignments have now been proposed for most of the well-known vinyl homopolymers. With this information, trends or internal consistencies among various assignments can be noted. At the same time, pitfalls that lead to erroneous conclusions can also be identified. For example, similarities do exist among the configurational assignments for methyl branched vinyl polymers such as poly(methyl methacrylate) and poly-(α-methylstyrene). However, methine assignments in vinyl homopolymers may occur as either *mm*, *mr*, *rr*, or *rr*, *mr*, *mm* from low to high field.

Although literature spectra are not reproduced, ^{13}C NMR spectra are presented of polymers from commercial and other sources when available for the convenience of the reader. The spectra were obtained in the author's laboratory under the same experimental conditions and can, therefore, be readily compared. The examples given demonstrate the broad range of applications of ^{13}C NMR but are not intended to be a comprehensive review of all commercial copolymers and homopolymers. As a final note, it is assumed that the reader is familiar with the basic concepts surrounding Markovian and Bernoullian statistical analyses (Chapter 4); for those who are not or wish a greater depth of understanding, references (4, Chapter II) and (41) are also recommended.

6.1 POLY(VINYL CHLORIDE)†

The ^{13}C NMR spectra of poly(vinyl chloride)s, prepared by free radical polymerization, show seven methine carbon resonances and five well-resolved methylene carbon resonances. These multiplicities have been attributed to differences in chemical shifts arising from monomer units in different configurational sequences by both Inoue *et al.* (68) and Carman *et al.* (67); however, different sets of assignments have been proposed. The more reasonable assignments are those of Carman, although none of the proposed assignments have been substantiated through specific reference polymers and model compounds. The seven methine carbon resonances, assigned by Carman, appear in order from low to high field as *rr*(two), *mr*(two), *mmmm*, *mmmr*, and *rmmr*. Splittings were observed for the *rr* and *mr* triads; however, more specific assignments were not possible. The methylene assignments, also proposed by Carman (46), are *rrr*, *rmr*, *rrm*, *mmr* + *mrm*, and *mmm* from low to high field. Triad and tetrad configurational distributions can, correspondingly, be obtained from the relative

† See refs. (46, 66–69).

^{13}C intensities of the methine and methylene carbon resonances. Unfortunately, a direct check for internal consistency between the results from the methylene and methine intensity data cannot be obtained because the *mmr* and *mrm* tetrad resonances were not resolved. A close Bernoullian fit, however, was noted for both sets of resonances with a P_m of 0.45. Calculated and observed triad and tetrad distributions are given in Table 6-1. A ^{13}C NMR spectrum of a similar free radical polymerized poly(vinyl chloride) is shown in Fig. 6.1. The statistical fit of Table 6-1 and an interpretation of the ^{13}C shielding in terms of *trans* and *gauche* interactions formed the primary basis for the Carman assignments.

Carman's assignments also lead to number-average sequence lengths of like configurations and of *meso* and *racemic* additions calculated independently from the triad and tetrad configurational distributions of Table

TABLE 6-1 Triad and Tetrad Comonomer Distributions for Free Radical Poly(vinyl chloride)[a]

	Relative intensity			
	Observed		Calculated ($P_m = 0.45$)	Chemical shift (ppm, TMS)
Triad	Methine resonances			
rr(2)	0.291	$r = 0.551$	0.303	57.15 57.06
mr(2)	0.520		0.496	56.29 56.13
mm(3)	0.188	$m = 0.448$	0.202	55.42 55.29 55.16
Tetrad	Methylene resonances			
rrr	0.161		0.166	47.74
rmr	0.146		0.130	47.35
rrm	0.282		0.272	47.01
(*mmr* + *mrm*)	0.320	*mmr* = 0.213[b] *mrm* = 0.107	0.334	46.33
mmm	0.092		0.091	45.59

[a] Obtained from ^{13}C NMR data at 25 MHz and calculated assuming a Bernoullian probability model. [C. J. Carman, *Macromolecules* **6,** 725 (1973). Reprinted with permission: copyright by the American Chemical Society.]

[b] If Bernoullian.

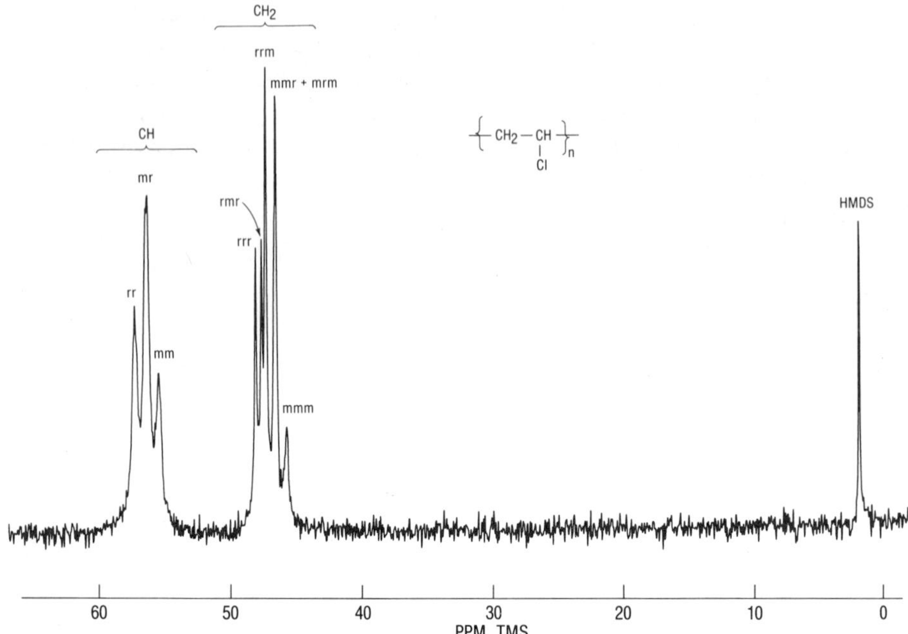

Fig. 6.1. Proton noise decoupled ^{13}C NMR spectrum at 25.2 MHz of a free radical initiated poly(vinyl chloride) in 1,2,4-trichlorobenzene at 120°C.

6-1. [See Eqs. (2.16), (2.29), and (2.40)–(2.43).] Overlap between the tetrads *mmr* and *mrm* can be resolved to $\frac{2}{3}(mmr + mrm)$ for *mmr* and $\frac{1}{3}(mmr + mrm)$ for *mrm* by assuming the polymer configurational distribution is Bernoullian. A number-average sequence length of 1.81 for like configurations is obtained independently from both the methine and methylene carbon intensity distributions. Correspondingly, a number-average sequence length of 1.82 is calculated from P_m using Eq. (2.17) expressed as

$$\bar{n} = 1/(1 - P_m) \tag{6.1}$$

Number-average sequences lengths calculated for both *meso* and *racemic* additions are 1.78 and 2.22, respectively, as obtained from the methylene carbon intensity distribution, and 1.72 and 2.12, respectively, as obtained from the methine carbon intensity distribution. These results, used collectively with those for like configurations, indicate the presence of an average configurational structure that is well punctuated with opposite additions such as

0 0 0 1 0 0 0 1 0 0 1 0 0 1 0 0 0 1 0
m m r r m m r r m r r m r r m m r r

The assignments proposed by Carman, therefore, are internally consistent and supported by calculations of number-average sequence lengths. These results are also consistent with those of Heatley and Bovey (70) where one can calculate a number-average sequence for like configurations of 1.75 from a pentad distribution measured in a 220 MHz of free radical polymerized poly(vinyl chloride).

6.2 POLY(VINYL ALCOHOL)†

A ^{13}C NMR configurational sensitivity is observed for both the methine and methylene carbons of poly(vinyl alcohol). A 25.2 MHz spectrum is reproduced in Fig. 6.2. The methine carbons produce a well-spaced triplet approximately 65 ppm from an internal tetramethylsilane (TMS) standard and exhibit enough fine structure to suggest a pentad chemical shift sensitivity (47). Four methylene resonances, which indicate at least a tetrad chemical shift sensitivity, are observed near 45 ppm in the ^{13}C NMR spectrum of poly(vinyl alcohol). Spectra can be obtained in either D_2O or DMSO-d_6 (47), although the latter solvent appears to give spectra with more resolved fine structure.

Different spectral assignments have been presented in independent studies by Wu and Ovenall (47) and Inoue *et al.* (71). Although neither set of assignments has been proven beyond any doubt, the assignments preferred are probably those of Wu and Ovenall. The monomer distributions from both the methine and methylene ^{13}C resonances are consistent with a triad distribution determined independently from hydroxyl resonances observed at 220 MHz. The assignments of Inoue *et al.* were not compared to the corresponding methine resonances but were based only on a Bernoullian fit. The results of Wu and Ovenall are not only consistent with Bernoullian behavior but are internally consistent as well.

Calculated and observed triad and tetrad distributions are given in Table 6-2 for poly(vinyl alcohol) as determined from both ^{13}C and ^{1}H NMR. The poly(vinyl alcohol) described in Table 6-2 was prepared by free radical polymerization and gives a comonomer distribution very similar to that previously observed for free radical poly(vinyl chloride). The *mm* versus *rr* methine assignment would have been open to question because of the small difference observed between these ^{13}C intensities. However, Wu and Ovenall were able to make unambiguous *mm* and *rr* assignments through an exam-

† See refs. (47, 71).

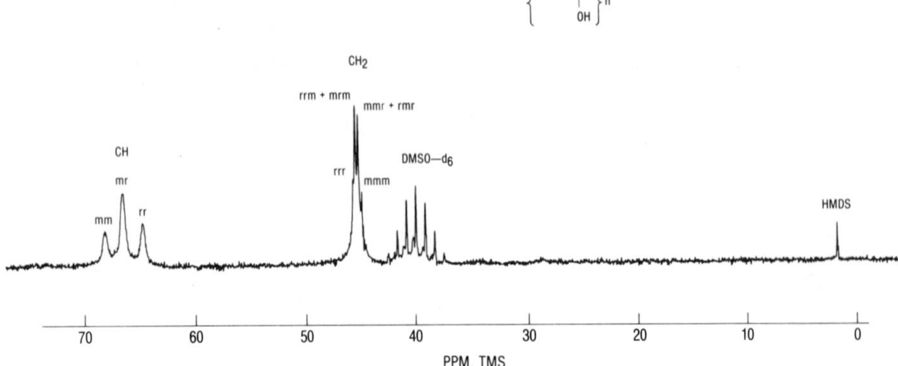

Fig. 6.2. Proton noise decoupled ^{13}C NMR spectrum at 25.2 MHz of poly(vinyl alcohol) in dimethylsulfoxide-d_6 at 100°C.

TABLE 6-2 Triad and Tetrad Comonomer Distributions for Poly(vinyl alcohol)[a]

	Relative intensity					
	Observed				Calculated	Chemical shift[b]
	^{13}C		^{1}H (220 MHz)		(P_m = 0.456)	(^{13}C ppm, TMS)
Triad	Methine resonances					
mm	0.23	*m* = 0.49	0.207	*m* = 0.456	0.208	67.8
mr	0.52		0.498		0.496	66.2
rr	0.25	*r* = 0.51	0.296	*r* = 0.545	0.296	64.3
Tetrad	Methylene resonances					
rrr	0.17	*r* = 0.54			0.161	45.8
rrm + *mrm*	0.37				0.383	45.6
mmr + *rmr*	0.35	*m* = 0.46			0.361	45.2
mmm	0.11				0.095	44.8

[a] Obtained from ^{13}C and ^{1}H NMR data and calculated assuming a Bernoullian probability model. [T. K. Wu and D. W. Ovenall, *Macromolecules* **6,** 582 (1973). Reprinted with permission; copyright by the American Chemical Society.]

[b] All chemical shifts are with respect to an internal TMS standard in $DMSO_4$-d_6.

ination of a cationic polymerized poly(vinyl alcohol) which was higher in *meso* additions. The methine assignments are critical because the order observed from low to high field for the triad is opposite to that reported for poly(vinyl chloride) (46). Triad and tetrad comonomer distributions for this second poly(vinyl alcohol) are listed in Table 6-3. The cationic poly-

TABLE 6-3 Triad and Tetrad Comonomer Distributions for a Cationic Polymerized Poly(vinyl alcohol)[a]

Triad	Observed ^{13}C	Observed ^{1}H	Tetrad	Observed ^{13}C NMR
mm	0.67	0.702	*rrr*	0.04
mr	0.25	0.225	(*rrm* + *mrm*)	0.14
rr	0.08	0.073	(*mmr* + *rmr*)	0.25
			mmm	0.57

[a] From T. K. Wu and D. W. Ovenall, *Macromolecules* **6,** 582 (1973). Reprinted with permission, copyright by the American Chemical Society.

(vinyl alcohol) comonomer distributions, however, cannot be fitted so well assuming a Bernoullian distribution. Number-average sequence lengths can be calculated independently of statistical fits and values of 1.83 and 5.39 were obtained for like configurations for the free radical and cationic polymers, respectively. These polymers had sufficiently different tacticities, the former possessing more *racemic* than *meso* dyads and the latter being predominantly *meso*, to provide a good test for assignments.

6.3 POLY(VINYL ACETATE)[†]

The ^{13}C NMR spectra of poly(vinyl actate)s (see Fig. 6.3) are not so well resolved as those from the corresponding poly(vinyl alcohol)s. In a study related to their earlier study of poly(vinyl alcohol), Wu and Ovenall (51) converted two previously well-characterized poly(vinyl alcohol)s, *meso* contents of 46 and 80%, respectively, to the corresponding poly(vinyl acetate)s. Acetylation of the alcohol groups was not expected to affect tacticity, thus polymers with known tacticities could be examined. Five resonances were detected for both the methylene and methine carbons; however, Wu and Ovenall were able to propose assignments for the methylene resonances only. The methine carbon resonances were not so well resolved and assignments would have been ambiguous. No configurational information could be obtained from the carbonyl or methyl carbon resonances because these resonances were apparently unresolved in the spectra obtained by Wu and Ovenall. A triplet can be observed for the carbonyl

[†] See refs. (20, 51, 72, and 83).

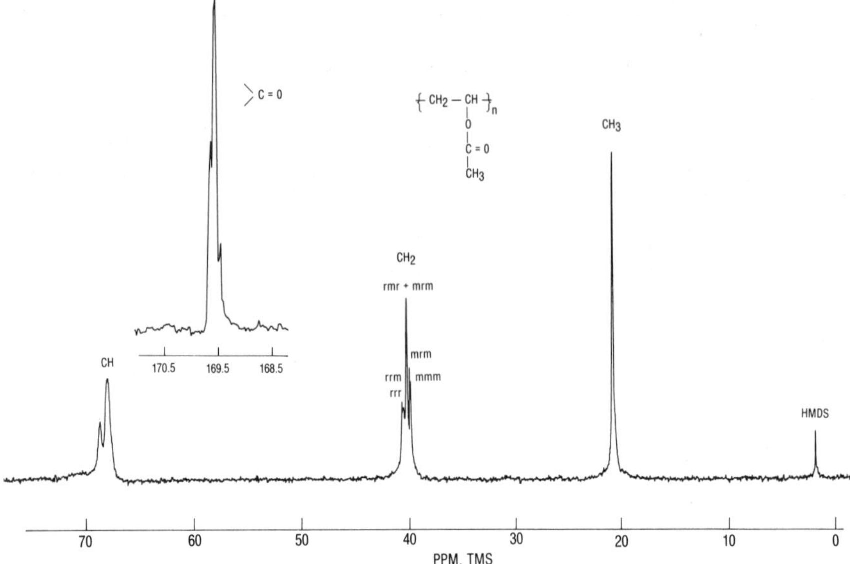

Fig. 6.3. Proton noise decoupled ^{13}C NMR spectrum at 25.2 MHz of poly(vinyl acetate) in 1,2,4-trichlorobenzene at 120°C.

TABLE 6-4 Tetrad Comonomer Distribution for Poly(vinyl acetate)[a]

Tetrad	Observed relative intensity	Poly(vinyl alcohol) distribution	Chemical shift (ppm, TMS)
	Methylene resonances ($\bar{n}$ = 1.8)		
rrr	0.20	0.17	38.2
rrm	0.25	0.25	37.8
rmr + *mrm*	0.25	0.28	37.7
mmr	0.17	0.19	37.6
mmm	0.13	0.11	37.3
	Methylene resonances ($\bar{n}$ = 5.4)		
rrr	0.0	0.04	38.2
rrm	0.1	0.06	37.8
rmr + *mrm*	0.1	0.08	37.7
mmr	0.2	0.25	37.6
mmm	0.6	0.57	37.3

[a] Obtained from ^{13}C NMR data and corresponding distribution from the poly(vinyl alcohol) precursor. [T. K. Wu and D. W. Ovenall, *Macromolecules* **7**, 776 (1974). Reprinted with permission, copyright by the American Chemical Society.]

resonance, however, as shown for the poly(vinyl acetate) polymer in Fig. 6.3. Methylene assignments of *rrr*, *rrm*, *rmr* + *mrm*, *mmr*, and *mmm* from low to high field gave triad distributions that were consistent with those observed in NMR analyses of the poly(vinyl alcohol) precursors; however, as shown in Table 6-4, only the assignments for *rrr*, *mmr*, and *mmm* are completely unambiguous. Unfortunately, the second and third resonances in the methylene multiplets have essentially the same intensity in both samples. Thus *rrm* and *rmr* + *mrm* can be interchanged without affecting the Bernoullian characteristics or the agreement among triad distributions from earlier poly(vinyl alcohol) analyses. Secondly, it is doubtful whether a cationic polymerized sample should obey Bernoullian statistics; thus the more isotactic sample may not be a good reference polymer for detailed assignments. Wu and Ovenall did select assignments that appeared reasonable and consistent with the earlier poly(vinyl alcohol) methylene assignments; however, final assignments should not be made until the internal consistency can be examined through methine carbon assignments.

6.4 POLY(METHYL METHACRYLATE)†

A ^{13}C NMR spectrum of poly(methyl methacrylate) is shown in Fig. 6.4. Splittings attributed to a sensitivity to configurational sequences are observed for all but the methoxy carbon, which exhibits a broad singlet. Peat and Reynolds (74) obtained self-consistent assignments through examination of two samples of substantially different tacticities. Final assignments were made by comparing observed intensity distributions for each carbon type with those calculated assuming both Bernoullian and Markovian statistical models. A well spaced triad is observed for the quaternary carbon while the α-methyl and side-chain carbonyl carbons exhibit essentially pentad chemical shift sensitivities. Listed in Table 6-5 are carbonyl, α-methyl, and quaternary carbon assignments and relative intensities that lead to the same triad distributions. These assignments are also consistent with a Bernoullian distribution having a P_m of 0.235.

The poly(methyl methacrylate) exhibiting Bernoullian characteristics was prepared by polymerization in tetrahydrofuran at −78°C with *n*-butyllithium as an initiator. A second polymer, prepared in toluene rather than tetrahydrofuran, was examined by Peat and Reynolds and showed either second-order Markov or Coleman–Fox (75) behavior. A fit was obtained

† See refs. (55, 74).

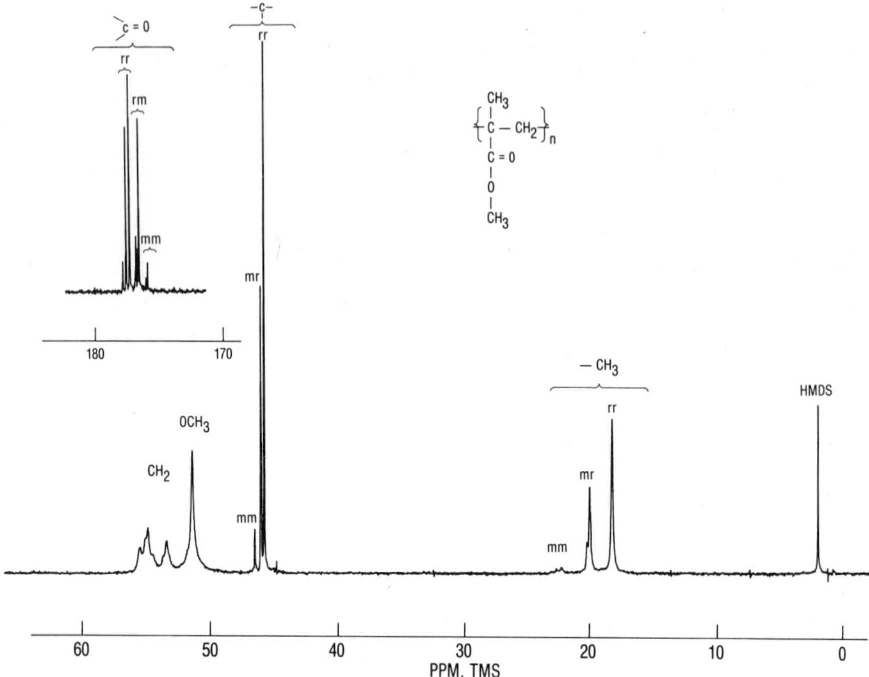

Fig. 6.4. Proton noise decoupled ^{13}C NMR spectrum at 25.2 MHz of poly(methyl methacrylate) in 1,2,4-trichlorobenzene at 120°C.

with the same assignments presented in Table 6-5, but applied to polymer that gave a different intensity distribution that strengthened the assignments.

The pentad chemical shift assignments for the carboxy carbon, from low to high field, are *mrrm*, *rrrm*, *rrrr*, *rmrm* + *mmrm*, *rmrr* + *mmrr*, *mmmm*, *mmmr*, and *rmmr*. These results are quite similar to the assignments proposed for the similarly situated quaternary carbon of poly(α-methylstyrene) where an assignment order of *mmrm*, *mrrr*, *rrrr*, *rmrr*, *mmrm*, *rmrm* + *mmrr*, and *rmmr* + *mmmr* + *mmmm* has been suggested (76). An interchange of *rmrr* and *rmrm* in either of these two vinyl polymers would lead to complete agreement between the assignments for the side-chain carbons. Based on

$$\left[-C(CH_3)(C(=O)OCH_3)-CH_2- \right]_n \qquad \left[-C(CH_3)(C_6H_5)-CH_2- \right]_n$$

TABLE 6-5 Triad and Pentad Comonomer Distribution for a Poly(methyl methacrylate) Conforming to Bernoullian Behavior[a]

Pentad	Triad	Carbonyl	Carbonyl	α-Methyl	Quaternary	Bernoullian (P_m = 0.235)	
						Pentad	Triad
mmmm		0.010				0.003	
mmmr	*mm*	0.018	0.051	0.051	0.060	0.020	0.005
rmmr		0.023				0.032	
mmrm		0.106				0.020	
rmrm						0.064	
	mr		0.367	0.374	0.361		0.359
mmrr		0.261				0.064	
rmrr						0.210	
mrrm		0.037				0.032	
mrrr	*rr*	0.191	0.582	0.575	0.579	0.210	0.586
rrrr		0.355				0.343	

[a] Obtained from ^{13}C NMR intensity distributions (74).

relative observed and predicted intensities, an interchange is possible; however, the relative intensities are too close to decide if any interchange should be made. The triad placements are *mm*, *mr*, and *rr* from low to high field for both the methyl and quaternary carbons. Once again, a similar situation occurred for poly(α-methylstyrene). It is interesting that these polymers, which possess such different side-chain groups, should show similarities in the order of configurational assignments.

Number-average sequence lengths for like configurations are 1.3 and 3.4 for the two poly(methyl methacrylate)s examined by Peat and Reynolds. Number-average sequence lengths for *meso* and *racemic* additions are 1.3 and 4.2, respectively, for the polymer exhibiting Bernoullian behavior ($\bar{n} = 1.3$). A predominantly syndiotactic structure, with only isolated *meso* additions, is indicated, that is,

$$\begin{matrix} r & r & r & m & r & r & r & m & r & m & r & r & r \\ 1 & 0 & 1 & 0 & 0 & 1 & 0 & 1 & 1 & 0 & 0 & 1 & 0 & 1 & 0 \end{matrix}$$

As observed from the triad distribution, one finds only 5% of the monomer units occurring in runs of three like configurations or longer.

6.5 POLY(α-METHYLSTYRENE)†

Poly(α-methylstyrene) exhibits strong chemical shift sensitivities toward configurational differences. In spectra taken at 67.88 and 90.51 MHz, Elgert and co-workers (76) performed a detailed study of three ^{13}C enriched polymers that were prepared by *n*-butyllithium initiation in tetrahydrofuran at temperatures ranging from +25°C to −78°C. As discussed in Section 6.4, the aromatic substituted carbon gives a spectral pattern and chemical shift sensitivity very similar to that observed for the carbonyl carbon in poly(methyl methacrylate). A basic chemical shift sensitivity to pentads is observed that yields the triad distribution *mm*:*mr*:*rr*. Triad information can also be obtained from each of the remaining ring carbons C_2, C_3, and C_4. The methylene carbons show a complex pattern that arises from a hexad sensitivity. Elgert *et al.* made detailed assignments for all except the *mmm*- and *mrm*-centered hexads. The quaternary carbon pattern was not as well resolved as the corresponding methylene and aromatic carbon resonances; however, assignments were made for the *rr*- and *mr*-centered pentads. In each of the regions examined for the various polymers, Elgert *et al.* found a

† See refs. (76, 77).

consistent conformity to Bernoullian behavior. Unfortunately, they did not choose to report the internal consistency of configurational distributions obtained from various spectral regions for a polymer system that may be the most amenable to such an analysis of any of the polymers discussed so far. A ^{13}C NMR spectrum of a poly(α-methylstyrene) obtained at 25.2 MHz is shown in Fig. 6.5. Although this spectrum is not as detailed as those taken at higher frequencies, it clearly shows the varied sensitivity to configurational sequences. The Bernoullian fit for the aromatic quaternary carbon is given in Table 6-6. The remaining fits, also given by Elgert and co-workers (76), were not identified according to the specific polymer examined and, consequently, cannot be used to prove internal consistency. The aromatic quaternary carbon gives the most interesting results because it shows a pentad sensitivity characteristic of side-chain groups, but gives chemical shifts that parallel those observed for the carbonyl carbons in poly(methyl methacrylate). The structure of the poly(α-methylstyrene)s reported by Elgert *et al.* also closely corresponds to that for the similarly prepared poly(methyl methacrylate) reported by Peat and Reynolds (74). Number-average sequence lengths of 1.3–1.5 for like configurations were obtained for the three poly(α-methylstyrene)s. This result demonstrates that these polymers are predominantly syndiotactic.

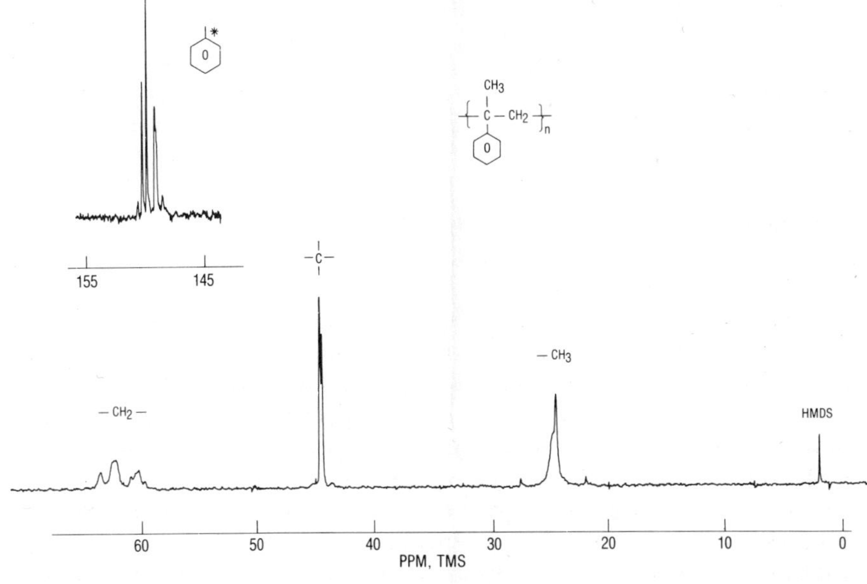

Fig. 6.5. Proton noise decoupled ^{13}C NMR spectrum at 25.2 MHz of poly(α-methylstyrene) in 1,2,4-trichlorobenzene at 120°C.

TABLE 6-6 Triad, Pentad Comonomer Distributions for a Poly(α-methylstyrene)[a]

		Observed relative		Calculated relative	
Pentad	Triad	Pentad	Triad	Pentad	Triad
mrrm		0.067		0.050	
mrrr	*rr*	0.207	0.452	0.196	0.438
rrrr		0.178		0.192	
rmrr		0.188		0.196	
mmrm	*mr*	0.047	0.460	0.052	0.448
rmrm		0.116		0.100	
mmrr		0.109		0.100	
	mm		0.089		0.114

[a] Determined from the aromatic quaternary carbon resonances. [K. F. Elgert *et al.*, *Polymer* **16,** 465 (1975), by permission of the publishers IPC Business Press Ltd. ©.]

One should not anticipate corresponding chemical shift behavior even in closely related polymers, because conformational properties associated with a particular configuration determine the chemical shifts. It is, therefore, interesting when such a correspondence does occur, as observed for poly(α-methylstyrene) and poly(methyl methacrylate). This result does suggest that similar shielding relationships exist between these two polymers. Whether or not this observation should be extended to suggest a similarity in steric relationships would require supporting evidence from some independent technique.

6.6 POLYSTYRENE†

The ^{13}C NMR spectra of polystyrenes prepared by free radical and *n*-butyllithium polymerizations show similar ^{13}C NMR spectral intensity distributions with nine distinct methylene resonances discernable and a single but broad methine resonance. A ^{13}C NMR spectrum of an *n*-butyllithium initiated polystyrene is shown in Fig. 6.6 and is similar to the free radical polystyrene in Fig. 4.1. A highly complex aromatic ring carbon resonance pattern is exhibited by the quaternary carbon with no splittings observed for either the ortho, meta, or para carbons. An expanded ^{13}C spectrum showing the aromatic quaternary carbon resonances is given in Fig. 6.7.

† See refs. (19, 55, 56, 77).

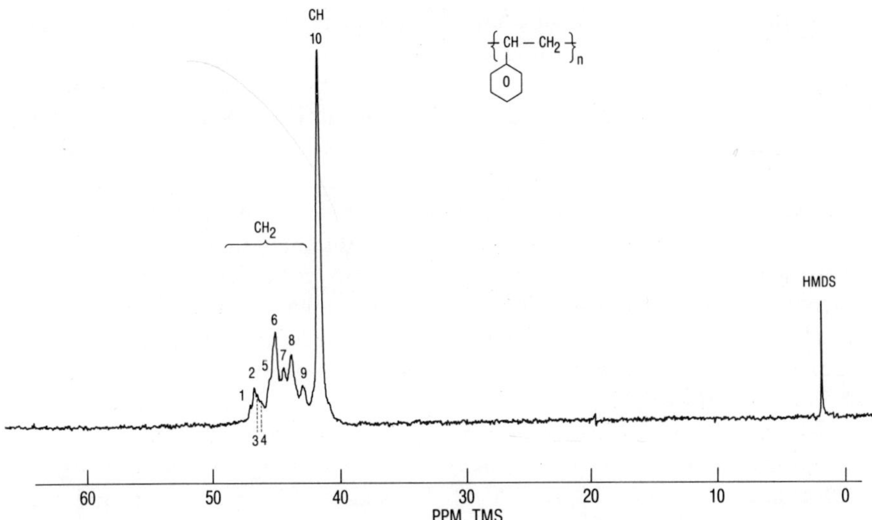

Fig. 6.6. Proton noise decoupled ^{13}C NMR spectrum at 25.2 MHz of a *sec*-butyllithium initiated polystyrene in 1,2,4-trichlorobenzene at 120°C.

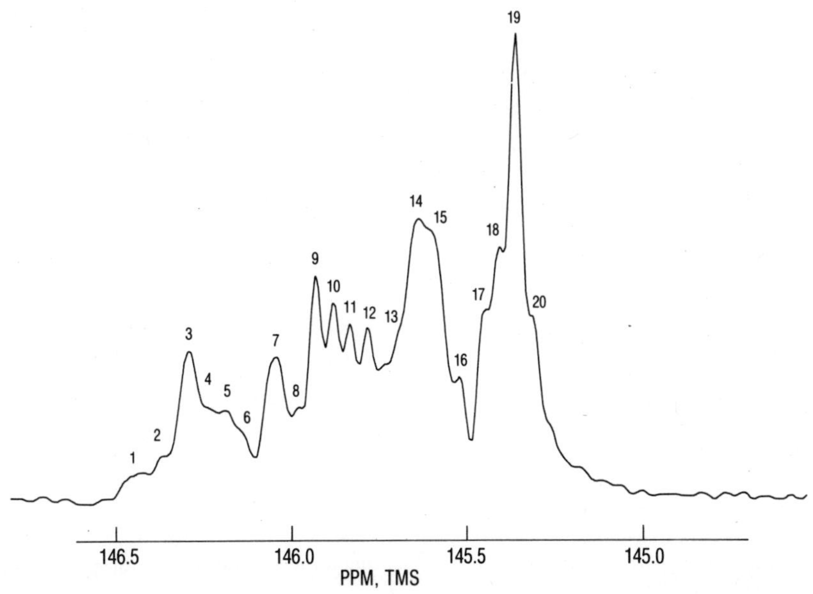

Fig. 6.7. Expanded view of the aromatic quaternary carbon resonances of the polystyrene shown in Fig. 6.6.

As discussed previously in Chapter 4, the methylene chemical shift sensitivity is basically tetrad with hexad splittings clearly visible for one tetrad sequence. An even higher sensitivity may be indicated for the other tetrads because of shoulders and broadening as shown in Fig. 6.8. Initial methylene ^{13}C chemical shift assignments, based upon one established *mmm* assignment and the supposition that free radical and alkyllithium polymerized polystyrenes are predominantly syndiotactic (54, 55), were made by Inoue and co-workers (77). The *mmm* assignment was made by comparison to a polystyrene containing predominantly isotactic sequences. (A similar polystyrene is shown in Fig. 4.2.) The methylene resonances were then assigned from low to high field to *rmr*, *mmr*, *rrr*, *mmr* + *mrr*, and *mrm*; however, the intensity distribution for these methylene assignments did not conform to Bernoullian statistics.

Aromatic quaternary carbon assignments were also made by Inoue *et al.* from low to high field: *mmmm*, *mmmr* + *rmmr*, *mmrm* + *rmrm*, *mmrr* + *rmrr*, *mrrm* + *mrrr*, and *rrrr*. Pentad assignments were chosen and the

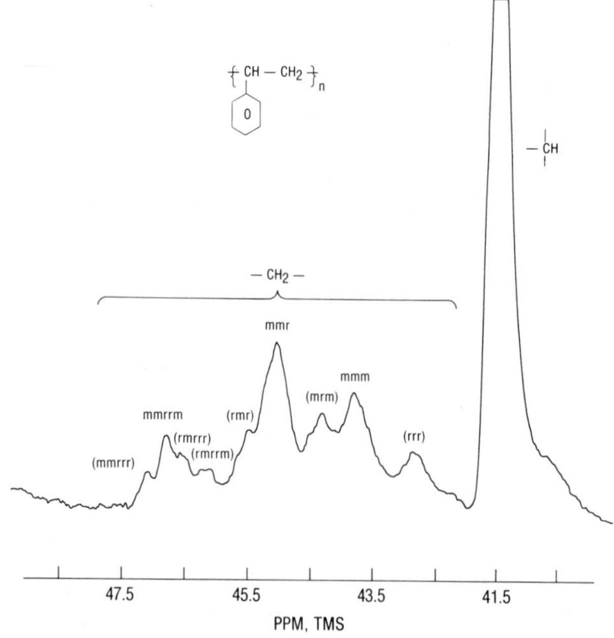

Fig. 6.8. Expanded view of the methylene and methine carbon resonances of the polystyrene shown in Fig. 6.6.

quaternary carbon resonances divided into five regions. The assignments were internally consistent because the tetrad–pentad relationships derived by Frisch *et al.* (7) were obeyed. Later, Matsuzaki *et al.* (56) examined a series of polystyrenes of different tacticities and made new quaternary carbon assignments of *mmmm* + *mmmr* + *rmmr* + *mrmm*, *rmrm* + *rrmm* + *mrrm*, and *rrmr* + *rrrm* + *rrrr*. The relative intensities of these quaternary carbon resonances were grouped into only three regions and found to conform to Bernoullian statistics. The aromatic quaternary carbon assignments proposed by Matsuzaki and associates were not well tested because relative area measurements were made of severely overlapping resonances. A Bernoullian fit was also obtained with only two independent observations and one independent variable. As seen in Fig. 6.7, the aromatic quaternary carbon region is a composite of perhaps as many as 20–25 closely spaced resonances.

The inconsistency in the Bernoullian behaviors of atactic polystyrenes reported by both Inoue and Matsuzaki and co-workers led to a reexamination of free radical and *n*-butyllithium polymerized polystyrenes (19). New methylene assignments, which were based on one "known" *mmm* assignment and a conformity to Bernoullian behavior, were presented from low to high field: *mmrrr*, *mmrrm*, *rmrrr*, *rmrrm*, *rmr*, *mmr*, *mrm*, *mmm*, and *rrr*. As discussed in Section 4.5, a Bernoullian probability was calculated from the *mmm* relative intensity and used to predict the remaining tetrad and hexad relative intensities. Close agreement was obtained between observed and calculated intensities. Assignments for *mmm*, *mmr*, and *mrr* could be easily discerned on a basis of the Bernoullian fit; however, the relative intensities of *rrr*, *mrm*, and *rmr* were too close for any but tentative assignments. These assignments are based on the assumption that the *mmm* tetrad has the same chemical shift in both the atactic and predominantly isotactic polymers. It is possible that the *mmm* tetrad has shifted in the spectrum of the atactic or amorphous polystyrene and that an identification of peak 8 is purely coincidental.

None of the ^{13}C spectral assignments, therefore, have been unambiguously proven for the configurational sequences in polystyrene. Model compounds (78) and reference polymers are needed. Epimerization experiments should be helpful for finalizing assignments for *mmm*, *mmr*, and *mrr*. Unfortunately, the same tetrads are produced whether a syndiotactic or isotactic polymer is epimerized. As discussed above, assignments based on strictly Bernoullian or Markovian behavior may not lead to conclusive results. More than one fit, which depends upon the manner in which a particular spectral region is assigned or divided, is possible.

6.7 POLY(METHYL ACRYLATE)†

A ^{13}C NMR spectrum obtained at 25.2 MHz of a poly(methyl acrylate) is shown in Fig. 6.9. Only the carboxy and methylene carbon resonances show splittings because of a sensitivity to configurational sequences. The methoxy and methine carbon resonances show little, if any, effects from polymer tacticity. Matsuzaki *et al.* (80) examined two poly(methyl acrylate)s, PMA-I and PMA-II, which had different tacticities, and interpreted the methylene resonances from a basis of a tetrad sensitivity. A Bernoullian fit, which was consistent with a configurational distribution measured independently from ^{1}H NMR, was obtained for both samples. Observed ^{1}H dyad and ^{13}C tetrad distributions and corresponding Bernoullian distributions are given in Table 6-7.

The availability of two samples with different tacticities strengthened the tetrad assignments because internally consistent results could be obtained. Unique assignments were not possible from PMA-I alone. By using values of P_m from ^{1}H NMR measurements, we limit the possible choices among Bernoullian fits and once again, strengthen the proposed assignments. The carboxy resonances were not assigned by Matsuzaki and co-workers, but were interpreted in terms of a pentad chemical shift sensitivity. Internally consistent assignments in the carboxy region would certainly further document the ^{13}C methylene assignments for the poly(methyl acrylate)s.

An independent calculation of the number-average sequence lengths in these poly(methyl acrylate)s depends upon resolution of each of the six methylene tetrads. Number-average sequence lengths based on the statistical fits are 2.0 and 3.7 for PMA-I and PMA-II, respectively. In this case, no real new information is obtained over that calculated from the tetrad distribution.

6.8 POLY(ISOPROPYL ACRYLATE)‡

Matsuzaki and associates (80) examined three poly(isopropyl acrylate)s with relative *meso* dyad concentrations of 0.35, 0.47, and 0.90. A configurational sensitivity was noted for the carboxy and backbone methylene and

† See refs. (79, 80).
‡ See ref. (80).

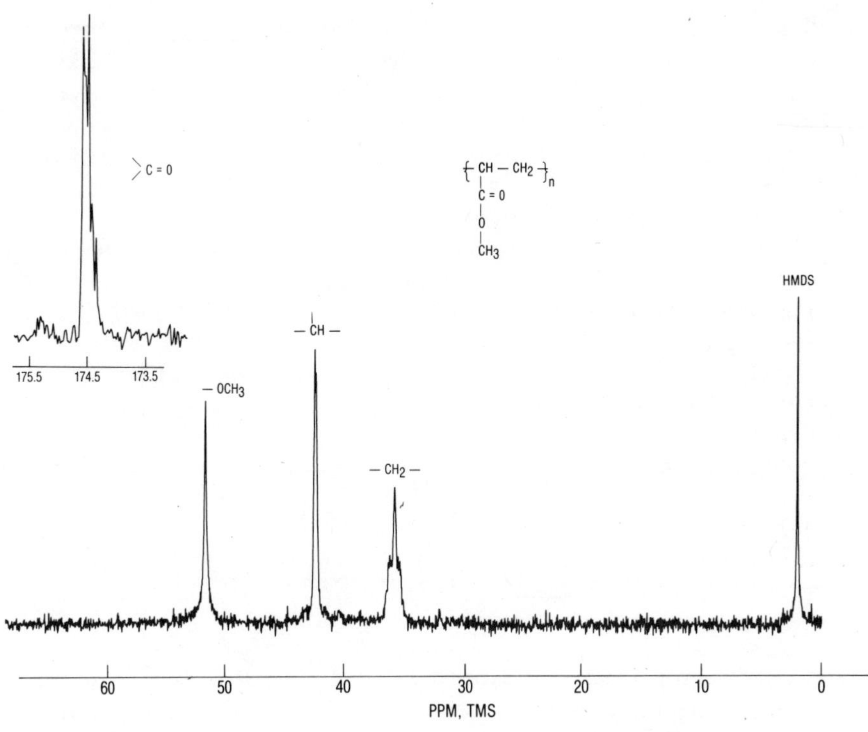

Fig. 6.9. Proton noise decoupled ^{13}C NMR spectrum at 25.2 MHz of poly(methyl acrylate) in 1,2,4-trichlorobenzene at 120°C.

TABLE 6-7 Calculated and Observed Tetrad Distributions[a]

	PMA-I			PMA-II		
Tetrad	Observed (CH_2) ($m = 0.49^b$)	Calculated ($P_m = 0.49$)		Observed (CH_2) ($m = 0.73^b$)	Calculated ($P_m = 0.73$)	
mmm	0.22	0.12	0.24	0.54	0.39	0.53
mrm		0.12			0.14	
mmr	0.52	0.25	0.50	0.39	0.29	0.40
mrr		0.25			0.11	
rmr	0.26	0.13	0.26	0.07	0.05	0.07
rrr		0.13			0.02	

[a] Measured from the methylene ^{13}C NMR intensity distributions of poly(methyl acrylate) (80).
[b] As measured from ^1H NMR.

methine carbons but did not occur for the side-chain methyl or methine carbons. A ^{13}C NMR spectrum of a poly(isopropyl acrylate) at 25.2 MHz is reproduced in Fig. 6.10. As in the analysis of the corresponding poly(methyl acrylate)s, a dyad distribution was obtained for each of the poly(isopropyl acrylate)s from ^{1}H NMR and used to calculate corresponding triad and tetrad distributions assuming Bernoullian behavior. The results are given in Table 6-8. The poly(isopropyl acrylate) with the highest *meso* dyad content was prepared by anionic polymerization and was not expected to conform to Bernoullian behavior. It was possible, however, to obtain a reasonably close first-order Markov fit as demonstrated by Matsuzaki and co-workers.

The methine resonances can be used to calculate, independently of any statistical fit, the number-average sequence length of like configurations and the number-average sequence lengths of both *meso* and *racemic* additions. For PiPA-I and PiPA-II, the following results were obtained:

	PiPA-I	PiPA-II
$\bar{n}$	1.5	1.9
$\bar{n}_m$	1.5	1.8
$\bar{n}_r$	3.1	2.1

The poly(isopropyl acrylate) labeled PiPA-II, prepared by free radical initiation, gives number-average sequence lengths for like configurations

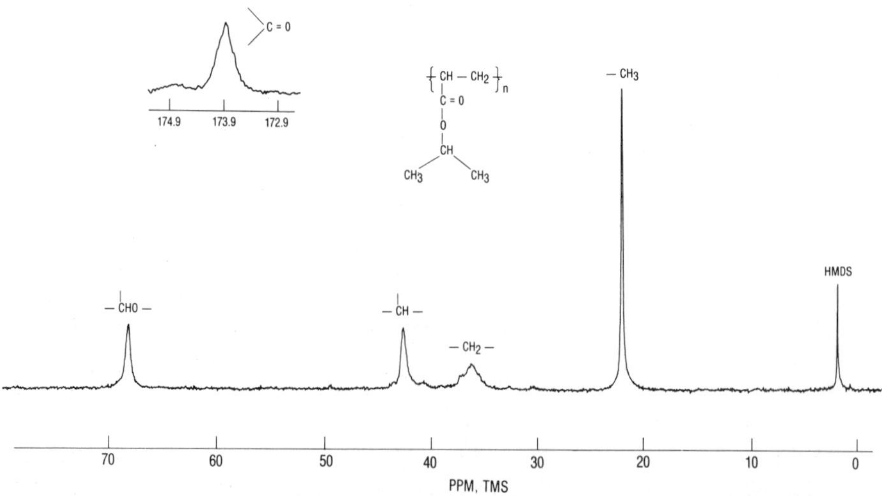

Fig. 6.10. Proton noise decoupled ^{13}C NMR spectrum at 25.2 MHz of poly(isopropyl acrylate) in 1,2,4-trichlorobenzene at 120°C.

TABLE 6-8 Calculated and Observed Triad and Tetrad Distributions[a]

	PiPA-I		PiPA-II		PiPA-III	
	Observed	Calculated ($P_m = 0.35$)[b]	Observed	Calculated ($P_m = 0.47$)[b]	Observed	Calculated ($P_m = 0.90$)[b]
Triad	Methine carbon resonances					
mm	0.10	0.12	0.21	0.22	0.85	0.81
mr	0.44	0.46	0.51	0.50	} 0.15	0.18 } 0.19
rr	0.46	0.42	0.28	0.28		0.01
Tetrad	Methylene carbon resonances					
mmm	0.03	0.04	0.09	0.10	0.81	0.73
mmr }		0.16		0.24		0.16
mrm }	0.50	0.08 } 0.54	0.60	0.12 } 0.62		0.08
mrr }		0.30		0.26	} 0.19	0.02 } 0.27
rmr }	0.47	0.15 } 0.42	0.31	0.13 } 0.28		0.01
rrr }		0.27		0.15		0.00

[a] Measured from the backbone methine and methylene ^{13}C NMR intensity distributions of poly(isopropyl acrylate) (80).

[b] As measured with ^{1}H NMR.

and for both *meso* and *racemic* additions of approximately 2.0 that is consistent with results expected for ideally random polymers. Number-average sequence lengths, independent of the statistical fit, cannot be determined for PiPA-III because the methine and methylene resonances were insufficiently resolved for complete triad or tetrad determinations.

6.9 POLY(METHYL VINYL ETHER)†

In a ^{13}C NMR spectrum of a poly(methyl vinyl ether) (81), the methoxy and methylene carbons exhibit multiplets from a stereochemical sensitivity while the methine carbon resonance shows no evidence of splittings. The methylene carbon resonances consist of two well-separated multiplets that can be readily used to measure relative *meso*, *racemic* dyad concentrations. The dyad assignments, agreed upon by both Johnson *et al* (55) and Matsuzaki *et al.* (81), ascribe the lower field multiplet to *racemic* dyads and the

† See refs. (55, 81).

upper field multiplet to *meso* dyads. Good agreement is obtained between the dyad distributions measured from ^{13}C relative intensities and calculated from a triad distribution obtained from ^{1}H NMR methoxy resonance data. Dyad distributions from both ^{1}H and ^{13}C NMR are given in Table 6-9 for two poly(methyl vinyl ether)s prepared by anionic polymerization. In neither case could a Bernoullian fit be obtained. As observed previously for ionically polymerized polymers (47, 74, 80). Bernoullian fits probably should not be expected.

The methoxy carbon resonances show more extensive splittings than do the methylene resonances and tentative pentad assignments have been made by Matsuzaki and associates from low to high field of *rrrr*, *mrrr* + *rmrr*, *mrrm* + *rmrm*, *mmrr*, *rmmr* + *mmmr* + *mmrm*, and *mmmm*. Qualitative agreement was noted between the pentad distributions of PMVE-1 and PMVE-2 based on these assignments and a first-order Markov distribution. A check for internal consistency between the methylene and methoxy carbon results is not possible until a complete pentad distribution can be obtained. Nevertheless, methylene dyad assignments are available that can be used to determine the tacticity of poly(methyl vinyl ether)s. Number-average sequence lengths for both like configurations and *meso*, *racemic* additions are

	PMVE-1	PMVE-2
$\bar{n}$	4.5, 4.8	2.9, 2.8
$\bar{n}_m$	5.8	3.3
$\bar{n}_r$	1.6	1.8

6.10 POLY(ETHYL VINYL ETHER)†

A ^{13}C NMR spectrum at 25.2 MHz of poly(ethyl vinyl ether) is reproduced in Fig. 6.11. Dyad splitting similar to that observed for poly(methyl vinyl ether) is observed for the backbone methylene carbons. Matsuzaki and co-workers used these methylene resonances to determine *racemic*, *meso* dyad concentrations in three different poly(ethyl vinyl ether)s. Results are given in Table 6-10 (81). Neither the backbone methine carbon nor ethoxy methyl carbon exhibited any configurational splittings. A triad observed for the methylene carbons, —CH_2O—, in the side-chain group could also be used to calculate relative dyad concentrations. The —CH_2O— triad resonances were assigned to *rr*, *mr*, and *mm* from low to high field. The

† See ref. (81).

TABLE 6-9 Dyad Distributions for Two Poly(methyl vinyl ether)s[a]

		PMVE-1			PMVE-2		
		^{1}H NMR Triad	Dyad[b]	^{12}C Dyad	^{1}H NMR Triad	Dyad[b]	^{23}C Dyad
mm	*m*	0.65	0.78	0.79	0.45	0.65	0.64
mr		0.27			0.40		
rr	*r*	0.08	0.22	0.21	0.15	0.35	0.36

[a] As measured from ^{1}H and ^{13}C NMR (81).
[b] The ^{1}H Dyad concentrations were determined by the necessary relationships, $(m) = (mm) + \frac{1}{2}(mr)$ and $(r) = (rr) + \frac{1}{2}(mr)$.

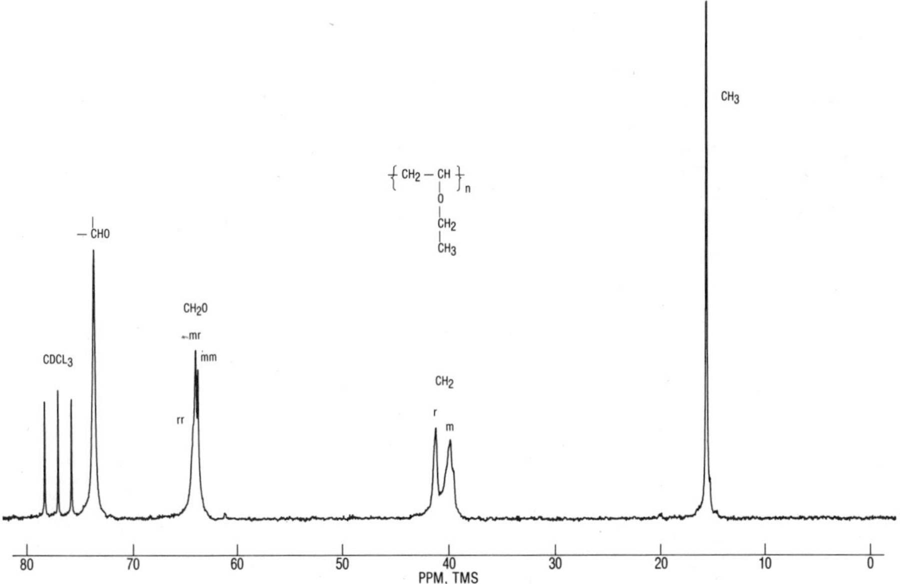

Fig. 6.11. Proton noise decoupled ^{13}C NMR spectrum at 25.2 MHz of poly(ethyl vinyl ether) in deuterochloroform at 48°C.

racemic methylene dyads from the polymer backbone were down field from the corresponding *meso* dyads as observed in poly(methyl vinyl ether). Further splittings were observed among these resonances but no assignments were made. As shown in Table 6-10, good agreement was obtained between dyad concentrations calculated from the —CH_2O— and —CH_2— resonances; thus the proposed assignments were internally consistent.

Each of these poly(ethyl vinyl ether)s PEVE-1, PEVE-2, and PEVE-3 were prepared by cationic polymerization but under reaction conditions that were

TABLE 6-10 *Meso, Racemic* Dyad Concentrations for Three Poly(ethyl vinyl ester)s of Varied Tacticity[a]

		PEVE-1			PEVE-2			PEVE-3		
		CH_2O		CH_2	CH_2O		CH_2	CH_2O		CH_2
		Triad	Dyad[b]	Dyad	Triad	Dyad[b]	Dyad	Triad	Dyad[b]	Dyad
mm	*m*	0.56	0.74	0.75	0.38	0.63	0.65	0.34	0.61	0.64
mr	*r*	0.36	0.26	0.25	0.50	0.37	0.35	0.54	0.39	0.36
rr		0.08			0.12			0.12		

[a] See Matsuzaki *et al.* (81).
[b] Dyad concentrations determined from the triad–dyad necessary relationships.

not the same. PEVE-1, prepared with a BF_3OEt_2 catalyst in toluene at $-78°C$, conforms reasonably well with a Bernoullian statistical model. The remaining polymers were prepared in methylene chloride at 0°C with BF_3OEt_2 used as the initiator for PEVE-2 and PF_5 used as an initiator for PEVE-3. Neither of these latter polymers conformed to Bernoullian behavior as can be seen from an inspection of the triad distributions given in Table 6-10. Unfortunately, a first-order Markov fit cannot be tested with only two independent observations and two variables.

Number-average sequence lengths, calculated from the respective dyad and triad distributions, are

	PEVE-1	PEVE-2	PEVE-3
$\bar{n}$	3.8, 4.0	2.7, 2.9	2.6, 2.8
$\bar{n}_m$	4.1	2.5	2.3
$\bar{n}_r$	1.4	1.5	1.4

Poly(ethyl vinyl ether)s 1 and 2 were prepared in the same manner as PMVE-1 and PMVE-2 discussed in the preceding section. For a comparison of the number-average sequence lengths, we find that shorter sequences of like configurations are favored for these ethoxy branched polymers.

6.11 POLY(ISOPROPYL VINYL ETHER) AND POLY(ISOBUTYL VINYL ETHER)[†]

Although apparent configurational sensitivities are shown by the backbone methylene carbons and both methine carbons of poly(isopropyl vinyl ether),

[†] See ref. (81).

Matsuzaki and co-workers (81) could not characterize the stereoregularity of a poly(isopropyl vinyl ether) prepared by PF_5 initiation in methylene chloride at 0°C. Well-separated backbone methylene dyad resonances, formerly observed for the corresponding poly(methyl vinyl ether)s and poly(ethyl vinyl ether)s, did not occur in the ^{13}C NMR spectrum of poly-(isopropyl vinyl ether). The methylene resonance was broad and, consequently, not further characterized. Three resonances were noted for the side-chain methine carbons; however, the distribution was so different from those of poly(methyl vinyl ether) and poly(ethyl vinyl ether) that Matsuzaki *et al.* chose not to interpret the resonances in terms of the simple triads *mm*, *mr*, and *rr*. Combinations of pentads with unlike triad centers to produce three resonances were considered to be the more probable situation.

A ^{13}C NMR spectrum of poly(isobutyl vinyl ether) is reproduced in Fig. 6.12. The only significant sensitivity toward configurational sequences occurs for the backbone methylene carbon resonances where two multiplets are observed analogous to poly(methyl vinyl ether) and poly(ethyl vinyl ether). Matsuzaki *et al.* assigned the lower field multiplet to *racemic* dyads and obtained a 38/62 *racemic* to *meso* distribution for a polymer prepared by PF_5 initiation in methylene chloride at 0°C. This distribution is similar to that obtained for the poly(ethyl vinyl ether) prepared in the same manner.

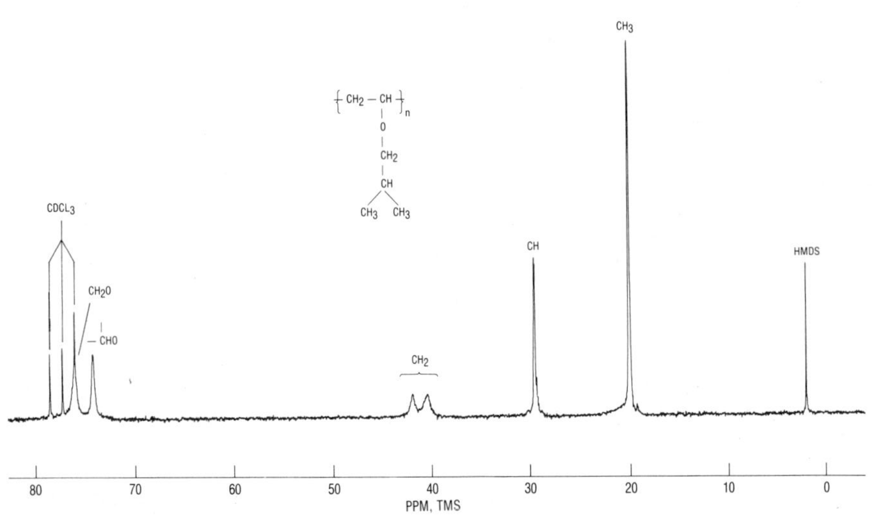

Fig. 6.12. Proton noise decoupled ^{13}C NMR spectrum at 25.2 MHz of poly(isobutyl vinyl ether) in deuterochloroform at 48°C.

6.12 POLY(*TERT*-BUTYL VINYL ETHER)†

In a ^{13}C NMR spectrum of poly(*tert*-butyl vinyl ether) (81), Matsuzaki and co-workers found that the methine and quaternary carbon resonances were the only sources of information about the tacticity of poly(*tert*-butyl vinyl ether)s. Resonances from the backbone methylene carbons did not show the dyad sensitivity characteristic of other polyvinyl ethers. Likewise, the methyl groups from the *tert*-butyl side-chain group displayed only a singlet resonance. Three resonances, which were assigned to *mm*, *mr*, and *rr* from low to high field, were observed for the pendant quaternary carbons. This direction is opposite to that assigned to pendant alkoxy carbons in poly(methyl vinyl ether) and poly(ethyl vinyl ether). The backbone methine carbons gave three major resonances with fine structure. A pentad distribution based on tentative methine carbon assignments, however, did correspond to the quaternary carbon triad distribution. The triad distribution measured from the quaternary carbon resonances was based on assignments that led to structures expected according to polymerization conditions; that is, the BF_3OEt_2 catalyst gives isotactic rich structures in nonpolar solvents and syndiotactic rich structures in polar solvents. The assignments presented by Matsuzaki *et al.* are therefore consistent with the trends established among various polyvinyl ethers; however, the assignments are considered tentative and more proof will be needed for final assignments. The polyvinyl ethers that give dyad splitting for the backbone methylene resonances, that is, poly(methyl vinyl ether), poly(ethyl vinyl ether), and poly(isobutyl vinyl ether), can be characterized because these assignments appear well established.

6.13 POLY(METHACRYLONITRILE)‡

A 25.2 MHz ^{13}C NMR spectrum of a poly(methacrylonitrile) is shown in Fig. 6.13. Splittings are observed for each carbon atom with the side-chain methyl group showing an apparent pentad chemical shift sensitivity. Assignments for four methyl resonances, given by Inoue and co-workers (82), are listed in Table 6-11. The *mr*-centered pentad assignments are considered tentative; however, the *mm*, *mr*, and *rr* triads appear to be identified and can be used to determine the tacticity of poly(methacrylonitrile). The triad

† See ref. (81).
‡ See ref. (82).

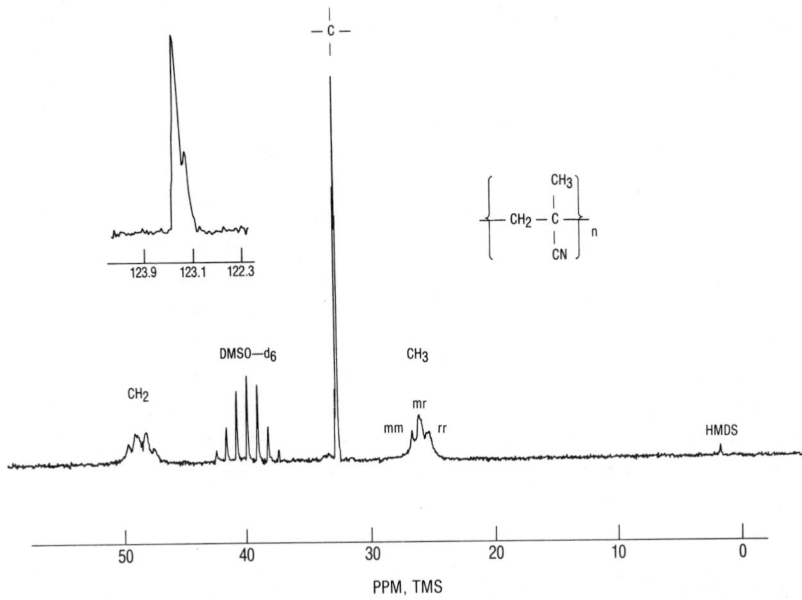

Fig. 6.13. Proton noise decoupled ^{13}C NMR spectrum at 25.2 MHz of poly(methacrylonitrile) in dimethylsulfoxide-d_6 at 100°C.

TABLE 6-11 Methyl ^{13}C NMR Triad and Pentad Assignments for Poly(methylacrylonitrile)[a]

Pentad	Triad	Relative intensity Observed		Relative intensity Calculated ($P_m = 0.4$)	
mmmm	*mm*	0.20		0.04	0.20
mmmr				0.10	
rmmr				0.06	
rmrr	*mr*	0.30	0.50	0.15	0.49
rmrm				0.12	
mmrr		0.20		0.12	
mmrm				0.10	
rrrr	*rr*	0.10	0.30	0.09	0.30
rrrm		0.14		0.15	
mrrm		0.06		0.06	

[a] Observed versus calculated intensity distribution. [Y. Inoue *et al.*, *Makromol. Chem.* **175,** 277 (1974), by permission of Hüthig and Weph Verlag, Basel.]

and pentad methyl assignments closely resemble those established for the methyl resonances in polypropylene. An interchange of *rmrm* with *mmrr* would lead to the same order as assigned in polypropylene for the side-chain methyl groups (22). It is not possible to distinguish the *mmrr* and *rmrm* pentads because the same relative intensity is obtained if the polymer conforms to Bernoullian behavior. Thus these assignments could be the same as those observed in polypropylene.

Inoue *et al.* examined a second poly(methacrylonitrile) that was prepared under reaction conditions designed to yield a polymer rich in isotactic dyads. The polymer methyl ^{13}C intensity distribution was non-Bernoullian. Assignments were made that were consistent with those listed in Table 6-11. A calculated distribution, which supported the proposed assignments, was obtained using the Coleman–Fox two-state model (75). As in the case of the free radical poly(methacrylonitrile), *rmrm* and *mmrr* could be interchanged to give an assignment "order" similar to that for polypropylene.

6.14 POLYACRYLONITRILE†

A ^{13}C NMR spectrum of a polyacrylonitrile is shown in Fig. 6.14. A well-defined configurational sensitivity is shown by the nitrile carbons that were used by Schaefer (83) to establish polymer tacticity. Although a pentad sensitivity is shown for the nitrile carbons, Schaefer only made tentative assignments according to triad centers *mm*, *mr*, and *rr* from low to high field. The *mm*- and *rr*-centered pentads are so well resolved in Fig. 6.14 that additional assignments could be made for *mmmr* and *mrrrr* after noting the Bernoullian characteristics of the intensity distribution.

Schaefer also noted a pentad chemical shift sensitivity for the methine carbon resonances with five of the ten pentads resolved. The methine resonances in Fig. 6.14 are not as well resolved as those obtained by Schaefer; however, three peaks are clearly visible. The methylene multiplet in Fig. 6.14 is a broad singlet with no distinct splittings. The polymer examined by Schaefer gave resonances that could be described as tetrad-sensitive; however, the multiplet intensities appeared to fall in an inverse order from those in the nitrile and methine multiplets.

Number-average sequence length measurements appear possible from the nitrile carbon resonances. From the relative intensities observed in Fig. 6.14, it appears unlikely that there is overlap among pentad resonances with dif-

† See ref. (83).

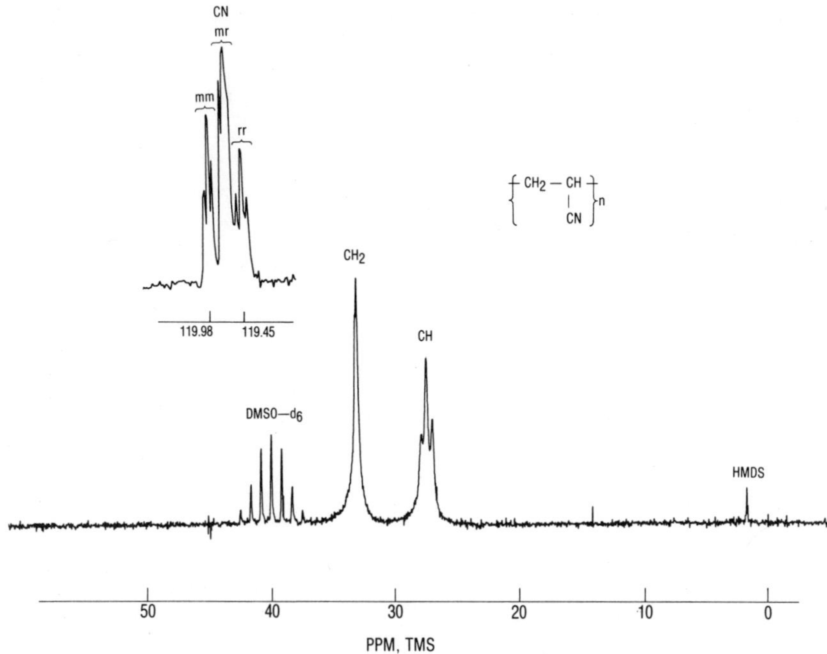

Fig. 6.14. Proton noise decoupled ^{13}C NMR spectrum at 25.2 MHz of polyacrylonitrile in dimethylsulfoxide-d_6 at 100°C.

ferent triad centers. However, as pointed out by Schaefer, overlap among pentads with different triad centers may be affecting the apparent methine carbon triad distribution. Although the nitrile pentad assignments have not been finalized, polyacrylonitrile is another example where the side-chain carbon resonance assignments are similar to those observed for the methyl groups in polypropylene.

6.15 TRENDS AMONG CONFIGURATIONAL ASSIGNMENTS IN VINYL HOMOPOLYMERS

Observed trends among chemical shift assignments for polymers could be useful for further assignments in ^{13}C NMR spectra if the validity of any trends could be established. As shown in Table 6-12, there is considerable assignment information associated with tacticity measurements in vinyl homopolymers. It is interesting to compile and compare assignments from

TABLE 6-12 Backbone Carbon Configurational Assignments in Vinyl Homopolymers[a]

Polymer	Methylene carbons	Methine carbons	Quaternary carbons
Poly(ethyl vinyl ether)	*r*	No splittings	
	m		
Poly(isobutyl vinyl ether)	*r*	No splittings	
	m		
Poly(isopropyl acrylate)	*rrr* + *rmr*	*rr*	
	mmr + *mrm* + *mrr*	*rm*	
	mmm	*mm*	
Poly(methyl acrylate)	*rrr* + *rmr*	No splittings	
	mmr + *mrr*		
	mmm + *mrm*		
Poly(methyl methacrylate)	No assignments		*mm*
			mr
			rr
Poly(α-methylstyrene)	*mrmrm*		*mrrm*
	rrmrm		*mrrr*
	rrmrr		*rrrr*
	mmmrr		*rmrr*
	mmmrm		*mmrm*
	rmmrr		*mmrr* + *rmrm*
	rmmrm		*mm*
	rrrrm		
Poly(α-methylstyrene)	*rrrrr*		
	mrrrm		
	mmm		
	mrm		
	rmrrm		
	rmrrr		
	mmrrr		
	mmrrm		
Poly(methyl vinyl ether)	*r*	No splittings	
	m		
Polypropylene	*mrm*	*mmmm*	
	rrr	*mmmr*	
	mrr	*rmmr* + *rrmm*	
	m	*mrmm* + *rrmr* + *mrmr* + *rrrr* + *mrrr* + *mrrm*	
Polystyrene	*mrr*	No splittings	
	rmr		
	mmr		
	mrm		
	mmm		
	rrr		
Poly(*tert*-butyl vinyl ether)	No splittings	Triad, pentad splittings	

TABLE 6-12 (*Continued*)

Polymer	Methylene carbons	Methine carbons	Quaternary carbons
Poly(vinyl acetate)	*rrr*	*mm*	
	rrm	No further assignments	
	mrm + *rmr*		
	mmr		
	mmm		
Poly(vinyl alcohol)	*rrr*	*mm*	
	rrm + *mrm*	*mr*	
	mmr + *rmr*	*rr*	
	mmm		
Poly(vinyl chloride)	*rrr*	*rr*	
	rmr	*mr*	
	rrm	*mmmr*	
	mmr + *mrm*	*rmmr*	
	mmm	*mmmm*	

[a] Assignments in the order of appearance from low to high magnetic fields.

various configurational sequences according to occurrence in the polymer spectrum. Chemical shift assignments for backbone methylene, methine, and quaternary carbons in order of appearance from low to high field are listed in Table 6-12 for polymers reviewed in this text. Although some of the assignments are considered tentative, all were listed for comparison purposes. It is interesting that no two polymers had the same methylene tetrad assignments. Generalizations concerning trends in the data should be approached cautiously; nevertheless, it can be noted that resonances from *racemic*-centered methylene sequences occurred at lower fields more frequently than did resonances from corresponding *meso*-centered sequences. Notable exceptions, however, are found among the assignments for polystyrene and poly(α-methylstyrene).

The methine carbon resonances, as a group, show the smallest chemical shift sensitivities toward configurational arrangements of the polymer carbons examined. Either no splitting or only triad splittings are observed in the majority of cases. In contrast, hexad splittings are frequently found among resonances for adjacent methylene carbons.

The side-chain carbon resonances have created the most interest and perhaps are used the most often as a source for tacticity measurements. Configurational assignments for side-chain carbons in vinyl homopolymers are listed in Table 6-13 in order of appearance from low to high field. The chemical shift sensitivity appears independent of the carbon type because

TABLE 6-13 Vinyl Polymer Side-Chain Carbon Configurational Assignments[a]

Polymer	Side-chain carbon
Polyacrylonitrile	—CN: *mm, mr, rr*
Poly(ethyl vinyl ether)	$—CH_2O$: *rr, mr, mm*
Poly(isopropyl acrylate)	C=O: No assignments; CH: No splittings; CH_3: No splittings
Poly(methacrylonitrile)	CH_3: *mmmm* + *mmmr* + *rmmr*, *rmrm* + *rmrr*, *mmrr* + *mmrm*, *rrrr*, *rrrm*, *mrrm*
Poly(methyl acrylate)	C=O: No assignments; OCH_3: No splittings
Poly(methyl methacrylate)	C=O: *mrrm*, *rrrm*, *rrrr*, *mmrm* + *rmrm*, *rmrr* + *mmrr*, *mmmm*, *mmr*, *rmmr* CH_3: *mm, mr, rr* OCH_3: No splittings
Poly(α-methylstyrene)	—C— :*mrrm*, *mrrr*, *rrrr*, *rmrr*, *mmrm*, *rmrm* + *mmrr*, *rmmr* + *mmmr* + *mmmm* C_o: *rr, mr, mm*; C_m: *rr, mr, mm*; C_p: *rr, mr, mm* CH_3: No splittings
Poly(methyl vinyl ether)	$CH_3O—$: *rrrr*, *mrrr* + *rmrr*, *mrrm* + *rmrm*, *mmrr*, *rmmr* + *mmmr* + *mmrm*, *mmmm*
Polypropylene	$—CH_3$: *mmmm*, *mmmr*, *rmmr*, *mmrr*, *mmrm*, *rmrr*, *rmrm*, *rrrr*, *mrrr*, *mrrm*
Polystyrene	Aromatic —C— : *mmmm*, no further assignments
Poly(*tert*-butyl vinyl ether)	—O—C— : *mm, mr, rr*
Poly(vinyl acetate)	C=O: No splittings $—CH_3$: No splittings

[a] Assignments in the order of occurrence from low to high magnetic fields.

the methyl carbon in polypropylene and the carbonyl in poly(methyl methacrylate) both exhibit pentad sensitivities. In Fig. 6.7, 20 resonances, suggesting at least a heptad chemical shift sensitivity, are observed for the quaternary carbon of the side-chain phenyl group of polystyrene. A consistency among side-chain carbon assignments is found for the carbonyl carbon of poly(methyl methacrylate) and the side-chain quaternary carbon of poly(α-methylstyrene). Similarities are also apparent for the side-chain methyl and nitrile carbon assignments for polypropylene, poly(methacrylonitrile), and polyacrylonitrile. No other trends among these polymer configurational assignments appear obvious.

In general, a tabulation of the chemical shift behavior among vinyl polymers demonstrates that assignments cannot be made by comparisons with related polymers. Chemical shifts are apparently determined by local screening effects, which may be optimized or minimized through conformational relationships among substituents possessing magnetic properties. Steric crowding does produce upfield chemical shifts (84); however, chemical shift behavior, which appears reasonable for one polymer system, should not be extended to explain the behavior observed in a second polymer. Even model compounds, which possess the same skeletal and substituent structure, may not reproduce the angular relationships found in polymers and thus should be used cautiously. Chemical shift assignments should therefore be reasoned independently for each vinyl polymer system examined.

6.16 ETHYLENE–PROPYLENE COPOLYMERS†

Ethylene–propylene copolymers were discussed in Chapter 3 as an example of a copolymer where head-to-tail monomer inversions were a necessary consideration in the ^{13}C NMR analysis. As shown in Figs. 3.4 and 6.15,

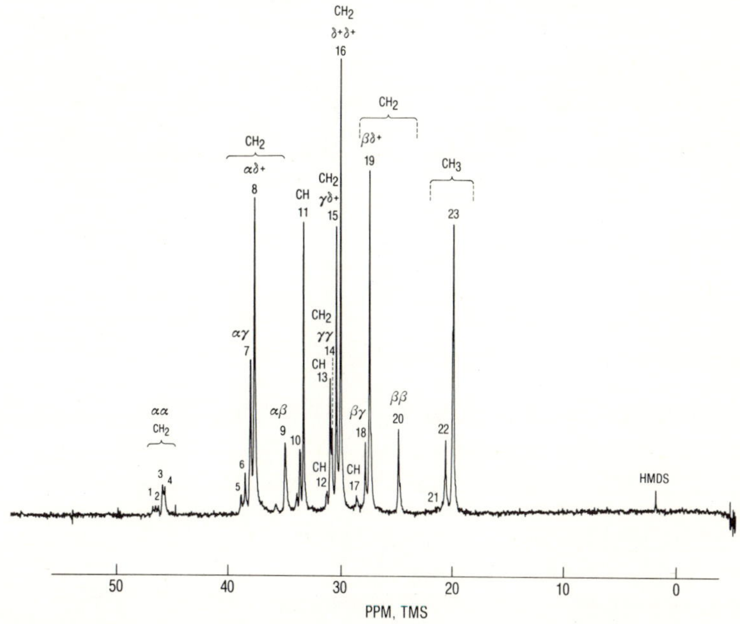

Fig. 6.15. Proton noise decoupled ^{13}C NMR spectrum at 25.2 MHz of a 50/50 ethylene–propylene copolymer in 1,2,4-trichlorobenzene at 120°C.

† See refs. (36–38, 85–89).

complex spectral patterns are obtained for ethylene–propylene copolymers. Initial assignments, based on calculated chemical shifts using the Grant and Paul parameters, were given by Crain and co-workers (36). Detailed assignments were later provided by Carman (37), who recognized that ethylene–propylene backbone carbon chemical shifts were determined largely by the proximity of methyl-branched carbons with lesser contributions from configurational arrangements. The methylene carbons discriminated by ^{13}C NMR and the respective chemical shifts are given below. (The Greek symbols suggested by Carman denote the locations of the nearest branch carbons.)

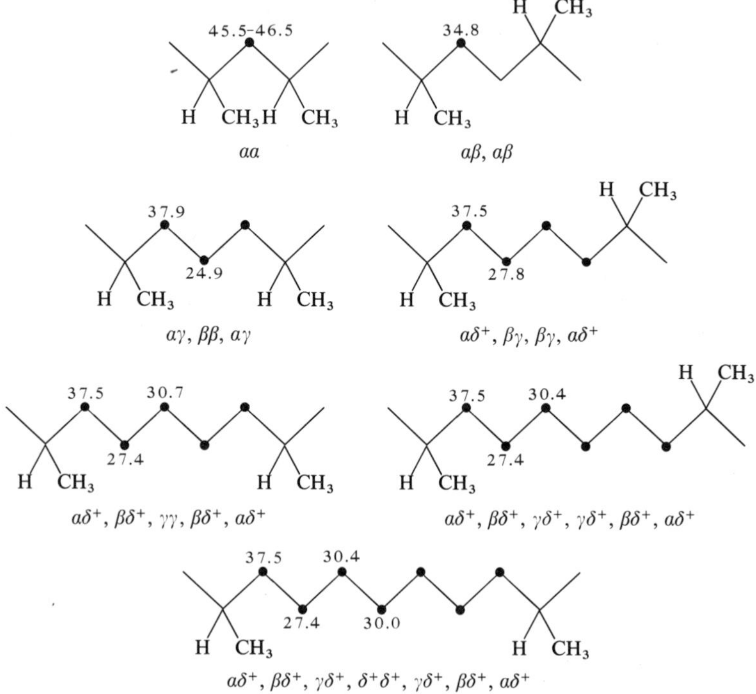

Methine and methyl carbon assignments, also from Carman (37), are

28.5
H CH3 H CH3 H CH3
21.3
PPP

30.9
31.2
H CH3 H CH3
20.7
EPP

33.2
H CH3
20.0
EPE

Although three distinct regions are evident for the methyl resonances in ^{13}C NMR spectra of ethylene–propylene copolymers, overlap from a corresponding difference in chemical shifts from configurational sequences and from ethylene–propylene arrangements prevent detailed assignments and unambiguous use in quantitative applications. For example, atactic polypropylenes (22) exhibit 10–12 methyl resonances over a range of chemical shifts from 19.8 to 21.7 ppm. This difference in chemical shifts is nearly the same as that shown above for the PPP, PPE, and EPE comonomer sequences. Also, as discussed in Chapter 1, an inversion of a propylene unit in PPP triads, of the type

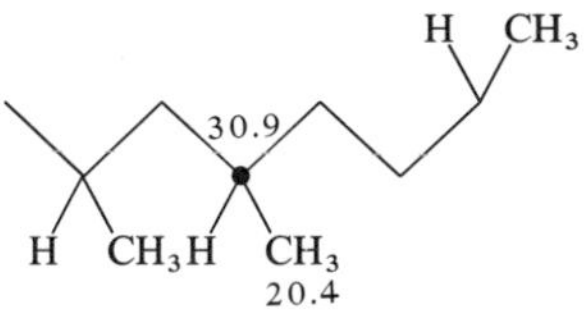

H CH_3

30.9

H CH_3 H CH_3

20.4

creates a methyl chemical shift that overlaps with those from PPE triads. Thus, the methyl region is difficult to analyze and may be of little use quantitatively unless a total methyl concentration is desired, or the examination is limited to polymers that contain no head-to-tail inversions and only isotactic propylene sequences (89).

Difficulties also exist in structural interpretations of methine carbon resonances although a lesser configurational sensitivity is present. The methine carbon resonances from PPP triads are only broadened by a configurational sensitivity while those from PPE triads show splittings that may be related to a pentad sensitivity. For reasons related to the methyl group behavior, the methine resonances in ethylene–propylene copolymers are also insensitive to propylene unit inversions. Thus the same chemical shifts occur for methine carbons in PPE triads as for PPP triads containing an inverted propylene sequence.

An opportunity for an excellent characterization of ethylene–propylene copolymers, however, is available from the methylene resonances. Successive methylene units from one to five in length give unique ^{13}C resonances. Those methylene carbons that are six or more carbons removed from a methine carbon give the same ^{13}C resonance. This chemical shift sensitivity closely parallels that observed for the methylene carbons in ethylene–1-olefin copolymers and low density polyethylenes (64, 87). Propylene unit inversions, therefore, can be identified unequivocally through the presence of methylene sequences two and four units in length. The presence of propylene unit inversions, however, precludes a determination of the three types of ethylene-centered triads, that is, EEE, PEE, and PEP. With inversions, the analysis may best be considered from a terpolymer viewpoint (38); however, there are insufficient ^{13}C NMR data to approach the problem in a manner similar to the hydrogenated butadiene–styrene copolymer of Section 3.7.

The most detailed study of ethylene–propylene copolymers has been presented by Carman *et al.* (38). Through a statistical study of ^{13}C NMR data from ethylene–propylene copolymers, they have obtained a complete analysis of the sequence distributions. Reactivity ratios r_1r_2 and percent propylene inversion were also important results from their statistical analysis of sequence distributions. Although Carman and co-workers emphasized a full use of ^{13}C intensity information, their most significant achievements were an extraction of monomer reactivity ratios from NMR data and a successful solution for the problem created by monomer inversion.

6.17 ETHYLENE–VINYL ACETATE COPOLYMERS†

Carbon-13 NMR data have been used to propose that the sequence distributions in ethylene–vinyl acetate copolymers are Bernoullian (20). Information concerning the comonomer distribution is available from the carbonyl, methine, and methylene regions. The acetate methyl group is apparently insensitive to both configurational and comonomer arrangements. More detailed assignments are given by Wu *et al.* (20) and Ibrahim *et al.* (73) than by Delfini *et al.* (91) because the latter only examined polymers with high ethylene contents. Wu's assignments are based on chemical shifts predicted with the Grant and Paul parameters augmented with acetate substituent effects calculated from a series of model compound; (20).

The carbonyl resonances are reported to be sensitive only to comonomer triad sequences with no splittings from configurational differences. The methylene and methine carbon chemical shifts are more difficult to interpret because a sensitivity to configurational differences is superimposed upon a comonomer sensitivity. Wu was able to predict the major line separations in both the methylene and methine regions with the Grant and Paul parameters and a comonomer distribution model that excluded vinyl acetate inversions. A triad sensitivity for the methylene carbon resonances and a mixed triad–pentad sensitivity for the methine carbon resonances were used satisfactorily to explain the observed ^{13}C spectral patterns. Smaller splittings, generally less than 1 ppm, were attributed to *meso*, *racemic* configurations. A ^{13}C NMR spectrum of a 60/40 ethylene–vinyl acetate copolymer is reproduced in Fig. 6.16.

† See refs. (20, 73, 91).

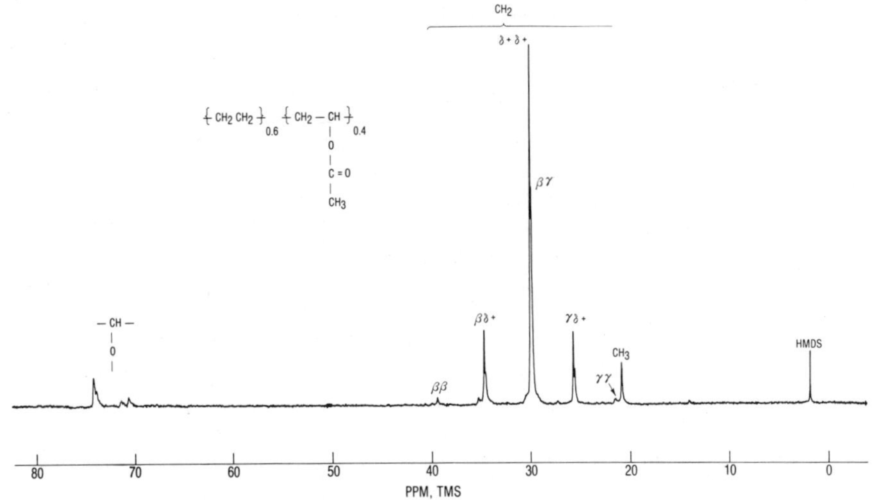

Fig. 6.16. Proton noise decoupled ^{13}C NMR spectrum at 25.2 MHz of a 60/40 ethylene–vinyl acetate copolymer in 1,2,4-trichlorobenzene at 120°C.

Relative concentrations of the vinyl acetate-centered triads VVV, VVE, and EVE obtained from ^{13}C methine and carbonyl resonances for the four copolymers examined by Wu and associated are listed in Table 6-14. Bernoullian distributions calculated from the ethylene and vinyl acetate mole fractions are included for comparison. Better agreement was obtained between the Bernoullian and observed methine triad distributions than between the Bernoullian and observed carbonyl triad distributions. This result probably occurred because the methine resonances were not so obscured by overlap as were the carbonyl resonances. A broadening of the carbonyl resonances, created by a configurational sensitivity, may have caused minor discrepancies among the relative intensities.

Number-average sequence lengths, calculated for vinyl acetate additions using the triad data in Table 6-14 and Eq. (3.14) are tabulated in Table 6-15. Corresponding results for ethylene additions could not be calculated because Wu and co-workers did not give concentrations for the ethylene-centered triads.

Number-average sequence lengths for the Bernoullian model in Table 6-15 were calculated with the observed vinyl acetate mole fractions and Eq. (4.13). These results quite naturally reaffirm the conclusions obtained from an inspection of the observed and calculated triad distributions. As the mole fractions of the vinyl acetate monomer decrease in the copolymer,

TABLE 6-14 Comonomer Triad Distributions for Ethylene–Vinyl Acetate Copolymers[a]

Mole fraction of VA		Comonomer triads EVE	VVE	VVV
0.735	Bernoullian	0.05	0.29	0.40
	Observed methine	0.07	0.30	0.37
	Observed carbonyl	0.06	0.32	0.36
0.48	Bernoullian	0.13	0.24	0.11
	Observed methine	0.12	0.26	0.11
	Observed carbonyl	0.09	0.24	0.14
0.34	Bernoullian	0.15	0.15	0.04
	Observed methine	0.15	0.15	0.04
	Observed carbonyl	0.12	0.17	0.05
0.10	Bernoullian	0.08	0.02	0.00
	Observed methine	0.09	0.01	0
	Observed carbonyl	0.07	0.03	0

[a] See Wu *et al.* (20).

TABLE 6-15 Number-Average Sequence Lengths for Vinyl Acetate Additions[a]

Mole fraction of VA	Source of triad data	Number-average sequence lengths: Observed	Bernoullian
0.735	Methine carbons	3.4	3.8
	Carbonyl carbons	3.4	
0.48	Methine carbons	2.0	1.9
	Carbonyl carbons	2.2	
0.34	Methine carbons	1.5	1.5
	Carbonyl carbons	1.7	
0.10	Methine carbons	1.1	1.1
	Carbonyl carbons	1.2	

[a] For a series of ethylene–vinyl acetate copolymers with vinyl acetate mole fractions of 0.735, 0.48, 0.34, and 0.10 (see Table 6-16).

we observe a better agreement between the observed sequence distributions and those calculated assuming a Bernoullian model.

Wu's assessment of the chemical shift differences resulting from carbon skeletal arrangements as opposed to configurational structures is characteristic of polymer ^{13}C NMR spectra. The former causes chemical shift differences of 5 to 40 ppm while *meso, racemic* configurations generally result in splittings of less than 1 ppm. Difficulties may arise when overlap obscures resonances for either configurational or comonomer arrangements, or inversion occurs for one of the monomer units. Wu and co-workers did not evaluate the possible effects of vinyl acetate inversion upon their ^{13}C NMR spectra.

Inversion of a vinyl acetate unit can create the same problems in ^{13}C NMR spectral analyses as found earlier in studies of the corresponding ethylene–propylene copolymers. For example, one cannot distinguish the following comonomer sequences from ^{13}C NMR spectra,

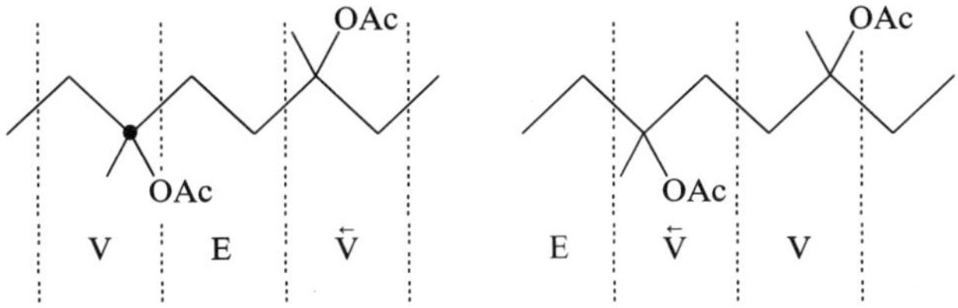

OAc OAc OAc OAc

V E $\overleftarrow{V}$ E $\overleftarrow{V}$ V

where $\overleftarrow{V}$ represents an "inverted" vinyl acetate sequence. As discussed in the analysis of ethylene–propylene copolymers (Sections 3.4 and 6.16), we can determine directly only the simple monomer distribution and a number-average sequence length of methylene carbons when inversion is present. Although Wu and Ibrahim were able to interpret their spectra satisfactorily without considering inversion, we may find it instructive for future reference to use the Grant and Paul parameters to calculate backbone carbon chemical shifts for each sequence possible in ethylene–vinyl acetate copolymers.

Acetate group parameter values, calculated through a linear regression analysis from data reported by Wu and co-workers (20), are given in Table 6-16. These values were used to calculate methylene chemical shifts associated with the following ethylene–vinyl acetate comonomeric sequences. [The

TABLE 6-16 Carbon-13 Substituent Parameters for the Acetate Group[a]

Parameter	Value (ppm)	Number of observations
α	50.4	12
β	6.0	15
γ_{CH_2OAc}	−6.2	5
γ_{CHOAc}	−4.2	6

[a] Corrective terms relating to differences in molecular geometry were apparently unnecessary for the acetate group.

notation is that suggested by Wu *et al.* (20), that is, the Greek letters denote the locations of the nearest acetate groups.]

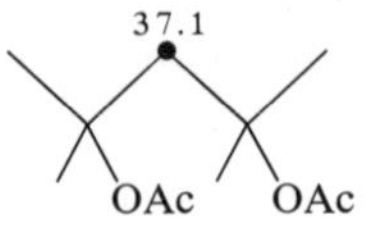

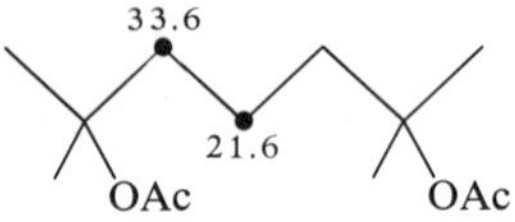

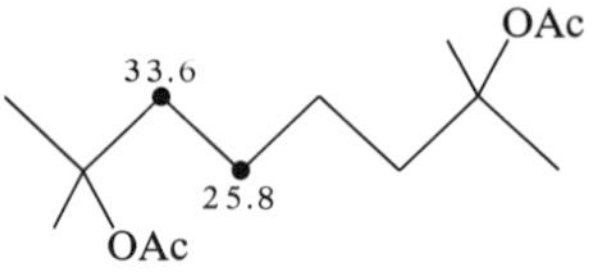

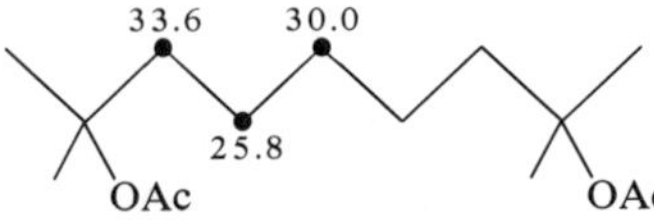

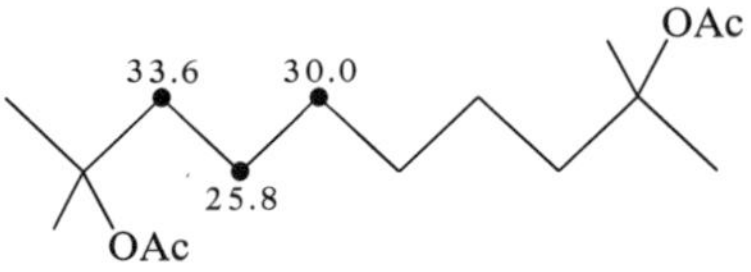

37.1
OAc OAc
$\beta\beta$

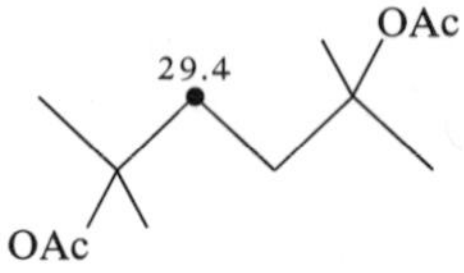

29.4
OAc
OAc
$\beta\gamma$

33.6
21.6
OAc OAc
$\beta\delta^+$, $\gamma\gamma$

33.6
25.8
OAc
OAc
$\beta\delta^+$, $\gamma\delta^+$

33.6 30.0
25.8
OAc OAc
$\beta\delta^+$, $\gamma\delta^+$, $\delta^+\delta^+$

33.6 30.0
25.8
OAc
OAc
$\beta\delta^+$, $\gamma\delta^+$, $\delta^+\delta^+$

There may be small chemical shift contributions from acetate groups four bonds away; it was necessary, however, to assume a negligible δ effect in the calculations of these chemical shifts. Values of δ obtained from linear regression were within experimental error of the total fit. Five methylene regions are predicted from the calculated results at 37, 34, 30, 26, and 22 ppm. These are in agreement with the observed methylene regions at 39, 34, 30, 25, and 21 ppm. An examination of the chemical shifts calculated for the above sequences leads to the conclusion that no discriminating chemical shifts are predicted for inverted vinyl acetate sequences. Different chemical shifts would occur if, in fact, a δ effect existed for the acetate group. In Fig. 6.16, the methylene resonances near 34, 30, and 26 ppm are split as expected for a δ acetate chemical shift effect and the existence of inverted vinyl acetate sequences. Splitting of the 30 and 26 ppm resonances was not reported by Wu *et al.*, thus inversion may not be present in their polymers. It is likely, however, that vinyl acetate inversion is present in the polymer shown in Fig. 6.16.

With respect to vinyl acetate inversions, the methine carbon regions in ^{13}C NMR spectra of ethylene–vinyl acetate copolymers are even more difficult to interpret than the methylene regions. Only two chemical shifts, which allow detection of monomer unit inversion, are predicted for the methine resonances according to the following methine shifts calculated with the Grant and Paul parameters:

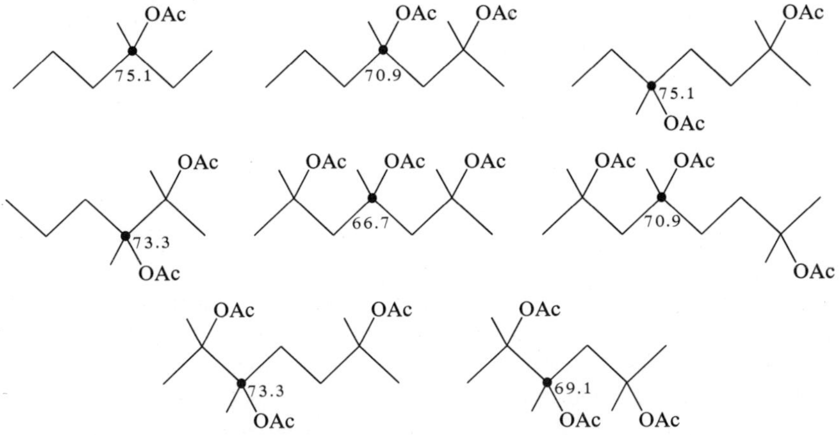

Three methine resonance regions were observed by Wu and co-workers at approximately 66, 70, and 75 ppm that were assigned to VVV, EVV, and EVE triads, respectively. Once again, inversion was not considered. The only type of inversion easily detected in the methine regions would be head-to-head vinyl acetate placements, which have methine carbon chemical shifts predicted at 69 and 73 ppm. The tail-to-tail vinyl acetate sequences do not give methine carbon chemical shifts independently of the sequences that contain no vinyl acetate inversion. A second factor complicating an interpretation of the methine carbon resonances would be an anticipated sensitivity toward configurational sequences. By assuming head-to-head and tail-to-tail placements to be absent, Wu *et al.* were able to offer an explanation of the observed methine resonance splittings on a basis of configurational and pentad comonomer placements.

Finally, the carbonyl resonances in the spectra of ethylene–vinyl acetate copolymers show a three-peak pattern that has been assigned by Wu *et al.* to triad comonomer sequences EVE, EVV, and VVV. Tacticity was not a consideration because the poly(vinyl acetate)s examined by Wu showed no effects from tacticity. Carbonyl splitting is observed, however, in the spectrum of the poly(vinyl acetate) shown in Fig. 6.3. The carbonyl carbon may be too far removed to give resonances that reflect head-to-tail inversions; thus the EVV and $\overleftarrow{V}\overleftarrow{V}V$ triad sequences and the EVE and $E\overleftarrow{V}V$ sequences will give the same resonances for the carbonyl group in the center of the triad. If head-to-tail inversions are known to be present, then the carbonyl triad distribution will not accurately reflect the simple VVV, VVE, and EVE distribution.

It would appear in the analyses of any copolymer involving ethylene and a vinyl monomer that careful attention should be given to the possibility of vinyl monomer inversion. As shown in Fig. 6.16, a $\beta\gamma$ resonance for inverted vinyl acetate sequences,

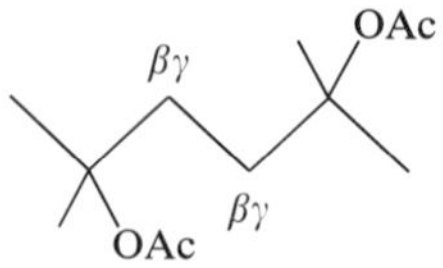

OAc
$\beta\gamma$
$\beta\gamma$
OAc

was detected in an ethylene–vinyl acetate copolymer. Such inversions can lead to inconsistencies in triad distributions from different regions. Any inconsistencies among various comonomer distributions should be carefully noted because this could be an additional means for detecting monomer unit inversions. Of course, synthetic routes that avoid inversion would be of great value in producing reference polymers for assignment purposes.

6.18 ACRYLONITRILE–STYRENE COPOLYMERS†

The monomer triad distribution for acrylonitrile–styrene copolymers was among the first ^{13}C NMR analyses of copolymers. The analysis was also one of the most difficult because two types of NMR chemical shift sensitivity were shown. Both monomer units have side-chain groups, nitrile and phenyl; thus a configurational sensitivity is superimposed upon the chemical shift sensitivity toward differences in neighboring monomer units. In spite of this complexity. Schaefer (92) was able to make enough ^{13}C assignments to obtain a complete triad monomer distribution. The analysis is based entirely on relative intensities obtained from the aromatic quaternary carbon resonances and nitrile resonances. Assignments and chemical shift behavior of the backbone methylene and methine carbons were not discussed.

Available to Schaefer were a series of reference acrylonitrile–styrene copolymers that were either blocked, alternating, or contained only isolated styrene additions. From these polymers, assignments could be made for each of the six unique triad combinations. The triad distribution for a low-conversion, nitrile-rich copolymer is given in Table 6-17.

† See ref. (92).

TABLE 6-17 Triad Comonomer Distribution for an Acrylonitrile–Styrene Copolymer[a]

Triad (0 = acrylonitrile, 1 = styrene)	Observed relative concentration
111	0.0
110	0.0
010	0.4
101	0.1
001	0.3
000	0.1

[a] Containing only "isolated" styrene additions. [J. Schaefer, *Macromolecules* **4,** 107 (1971). Reprinted with permission, copyright by the American Chemical Society.]

From Eqs. (3.14) and (3.5), number-average sequence lengths can be calculated for the styrene and acrylonitrile additions. A value of 2.0 is obtained for acrylonitrile and a value of 1.0, as expected, for styrene. The comonomer distributions, however, are only in fair agreement with the reported 64 mole % acrylonitrile. The triad distribution in Table 6-17 indicates a 60:40 acrylonitrile to styrene ratio.

6.19 BUTADIENE–STYRENE COPOLYMERS†

One of the most difficult ^{13}C NMR analyses is found in structure determinations of butadiene–styrene copolymers. There are three butadiene modes of addition possible in a butadiene–styrene copolymer: *cis*-1,4-butadiene, *trans*-1,4-butadiene, and 1,2-butadiene. When these butadiene additions are considered with styrene, we find a total of ten unique dyads and 40 triads. In a ^{13}C NMR analysis, we must establish whether the observed resonances originate from either dyad or triad comonomer sequences. Fortuitous overlap and chemical shift degeneracies must also be considered. In a detailed analysis of chemical shift behavior, Katritzky and Weiss (93, 94) have given 26 methylene and methine carbon assignments for butadiene–styrene copolymers.

Katritzky and Weiss made assignments based on the ^{13}C spectral behavior of the corresponding homopolymers and an assumed dyad chemical shift

† See refs. (93, 94).

sensitivity. These assignments appear reasonable overall and were used to obtain structural information about sequence distributions and number-average sequence lengths. A possible triad sensitivity for the styrene methine carbons, however, may have been overlooked. Three styrene methine resonances were observed in corresponding spectra of the hydrogenated butadiene–styrene copolymers (39), which were

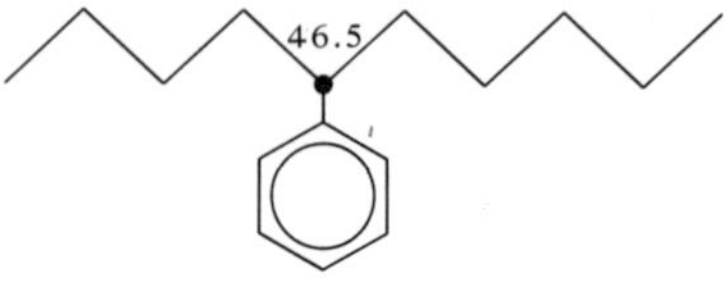

46.5

I

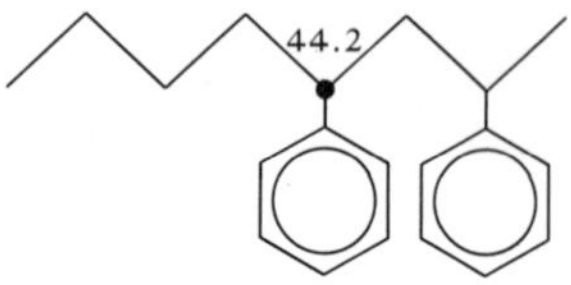

44.2

II

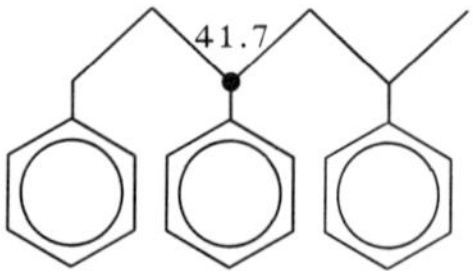

41.7

III

Similar methine chemical shifts are anticipated in the ^{13}C NMR spectra of butadiene–styrene copolymers because double bonds have a negligible chemical shift contribution when located β to the carbon of interest (95; 21, p. 59). Katritzky and Weiss correctly noted styrene methine carbon resonances at 47.2 and 41.7 ppm but assigned an observed resonance at 44.12 ppm only to methylene carbons in 1,2-butadiene–styrene sequences. We should also consider that the methine carbon resonance shown by structure II occurs at this position. Dyad concentrations required by Katritzky and Weiss can then be obtained from the triad concentrations of I, II, and III using Eqs. (3.26) and (3.30).

Katritzky and Weiss (96) have utilized ^{13}C NMR extensively to characterize sequence distributions in copolymers. Other studies of interest are the structural distributions found in butadiene–acrylonitrile (97) and styrene–methyl methacrylate (98) copolymers.

References

1. T. C. Farrar, *Anal. Chem.* **42**(4), 109A (1970).
2. T. C. Farrar and E. D. Becker, "Pulse and Fourier Transfer NMR." Academic Press, New York, 1971.
3. G. Natta and F. Danusso, *J. Polym. Sci.* **34,** 3 (1959).
4. F. A. Bovey, "Polymer Conformation and Configuration." Academic Press, New York, 1969.
5. H. L. Frisch, C. Schuerch, and M. Szwarc, *J. Polym. Sci.* **11,** 559 (1953).
6. F. P. Price, *J. Chem. Phys.* **36,** 209 (1962).
7. H. L. Frisch, C. L. Mallows, and F. A. Bovey, *J. Chem. Phys.* **45,** 1565 (1966).
8. H. L. Frisch, C. L. Mallows, F. Heatley, and F. A. Bovey, *Macromolecules* **1,** 533 (1968).
9. F. A. Bovey, *Acct. Chem. Res.* **1,** 175 (1968).
10. F. Heatley, R. Salovey, and F. A. Bovey, *Macromolecules* **2,** 619 (1969).
11. F. C. Stehling, *J. Polym. Sci. Polym. Chem. Ed.* **2,** 1815 (1964).
12. A. L. Segre, *Macromolecules* **1,** 93 (1968).
13. F. Heatley and A. Zambelli, *Macromolecules* **2,** 618 (1969).
14. D. M. Grant and E. G. Paul, *J. Amer. Chem. Soc.* **86,** 2984 (1964).
15. L. P. Lindeman and J. Q. Adams, *Anal. Chem.* **43,** 1245 (1971).
16. J. D. Roberts, "Nuclear Magnetic Resonance: Applications to Organic Chemistry," Chapter 3. McGraw-Hill, New York, 1959.

17. J. C. Randall, *J. Polym. Sci. Polym. Phys. Ed.* **13,** 901 (1975).
18. C. J. Carman, A. R. Tarpley, Jr., and J. H. Goldstein, *Macromolecules* **6,** 719 (1973).
19. J. C. Randall, *J. Polym. Sci. Polym. Phys. Ed.* **13,** 889 (1975).
20. T. K. Wu, D. W. Ovenall, and G. S. Reddy, *J. Polym. Sci. Polym. Phys. Ed.* **12,** 901 (1974).
21. G. C. Levy and G. L. Nelson, "Carbon-13 Nuclear Magnetic Resonance for Organic Chemists," p. 47. Wiley, New York, 1972.
22. J. C. Randall, *J. Polym. Sci. Polym. Phys. Ed.* **12,** 703 (1974).
23. G. Natta, *J. Polym. Sci.* **34,** 531 (1959).
24. F. C. Stehling and J. R. Knox, *Macromolecules* **8,** 595 (1975).
25. G. J. Listner, U.S. Patent 3,511,824, 1970.
26. A. Zambelli, P. Locatelli, G. Bajo, and F. A. Bovey, *Macromolecules* **8,** 687 (1975).
26a. A. Provasoli and D. R. Ferro, *Macromolecules* **10,** 874 (1977).
27. J. C. Randall, *J. Polym. Sci. Polym. Phys. Ed.* **14,** 2083 (1976).
28. D. Doddrell, V. Glushko, and A. Allerhand, *J. Chem. Phys.* **56,** 3683 (1972).
29. K. Kuhlmann, D. M. Grant, and R. K. Harris, *J. Chem. Phys.* **52,** 3439 (1970).
30. J. Schaefer and D. F. S. Natusch, *Macromolecules* **5,** 416 (1972).
31. H. L. Hsieh and W. H. Glaze, *Rub. Chem. Tech.* **43**(1), 22 (1970).
32. A. D. H. Clague, J. A. M. van Broekhoven, and L. P. Blaauw, *Macromolecules* **7,** 348 (1974).
33. J. C. Randall, *J. Polym. Sci. Polym. Phys. Ed.* **13,** 1975 (1975).
34. R. L. Arnett and C. J. Stacey, *J. Phys. Chem.* **77,** 1986 (1973).
35. E. R. Santee, Jr., R. Chang, and M. Morton, *J. Polym. Sci. Polym. Lett. Ed.* **11,** 449 (1973).
36. W. O. Crain, Jr., A. Zambelli, and J. D. Roberts, *Macromolecules* **4,** 330 (1971).
37. C. J. Carman and C. E. Wilkes, *Rub. Chem. Tech.* **44,** 781 (1971).
38. C. J. Carman, R. A. Harrington, and C. E. Wilkes, *Macromolecules* **10,** 536 (1977).
39. J. C. Randall, *J. Polym. Sci. Polym. Phys. Ed.* **15,** 1451 (1977).
40. F. E. Naylor, H. L. Hsieh, and J. C. Randall, *Macromolecules* **3,** 486 (1970).
41. F. P. Price, "Markov Chains and Monte Carlo Calculations in Polymer Science" (G. G. Lowry, ed.). Dekker, New York, 1970.
42. T. Alfrey, Jr., and G. Goldfinger, *J. Chem. Phys.* **12,** 205 (1944).
43. E. R. Santee, Jr., V. D. Mochel, and M. Morton, *J. Polym. Sci. Polym. Lett. Ed.* **11,** 453 (1973).
44. K. Hatada, Y. Tanaka, Y. Terawaki, and H. Okuda, *Polym. J.* **5,** 327 (1973).
45. Y. Tanaka, H. Sato, M. Ogawa, K. Hatada, and Y. Terawaki, *J. Polym. Sci. Polym. Lett. Ed.* **12,** 369 (1974).
46. C. J. Carman, *Macromolecules* **6,** 725 (1973).
47. T. K. Wu and D. W. Ovenall, *Macromolecules* **6,** 582 (1973).
48. F. A. Bovey and G. V. D. Tiers, *J. Polym. Sci.* **44,** 173 (1960).
49. B. D. Coleman, *J. Polym. Sci.* **31,** 155 (1968).
50. J. R. Woodbrey, "The Stereochemistry of Macromolecules" (A. D. Kelley, ed.), Vol. 3, Chapter 2. Dekker, New York, 1968.
51. T. K. Wu and D. W. Ovenall, *Macromolecules* **7,** 776 (1974).
52. J. R. Ebdon and T. N. Huckerby, *Polymer* **17,** 170 (1976).
53. K. Matsuzaki, T. Uryu, K. Osada, and T. Kawamura, *J. Polym. Sci. Polym. Chem. Ed.* **12,** 2873 (1974).
54. S. Brownstein, S. Bywater, and D. J. Worsfold, *J. Phys. Chem.* **66,** 2067 (1962).
55. L. F. Johnson, F. Heatley, and F. A. Bovey, *Macromolecules* **3,** 175 (1970).
56. K. Matsuzaki, T. Uryu, K. Osada, and T. Kawamura, *Macromolecules* **5,** 816 (1972).
57. R. S. Porter, M. J. Cantow, and J. F. Johnson, *Makromol. Chem.* **94,** 143 (1966).
58. Y. Inoue, A. Nishioka, and R. Chûjô, *Makromol. Chem.* **168,** 163 (1973).
59. J. C. Randall, *J. Polym. Sci. Polym. Phys. Ed.* **14,** 1693 (1976).

60. R. Freeman, K. G. R. Pachler, and G. N. La Mar, *J. Chem. Phys.* **55,** 4586 (1971).
61. G. C. Levy, U. Edlund, and J. G. Hexem, *J. Magn. Resonance* **19,** 259 (1975).
62. E. M. Galimov, NASA TT-F-682 (1975).
63. D. J. Frank and W. M. Sackett, *Geochim. Cosmochim. Acta.* **33,** 811 (1969).
64. J. C. Randall, *J. Polym. Sci. Polym. Phys. Ed.* **11,** 275 (1973).
65. J. A. Pople, W. G. Schneider, and H. J. Bernstein, "High-resolution Nuclear Magnetic Resonance," p. 37. McGraw-Hill, New York, 1959.
66. J. Schaefer, *Macromolecules* **4,** 110 (1971).
67. C. J. Carman, A. R. Tarpley, Jr., and J. H. Goldstein, *Macromolecules* **4,** 445 (1971).
68. Y. Inoue, I. Ando, and A. Nishioka, *Polym. J.* **3,** 246 (1972).
69. C. J. Carman, A. R. Tarpley, Jr., and J. H. Goldstein, *J. Amer. Chem. Soc.* **93,** 2864 (1971).
70. F. Heatley and F. A. Bovey, *Macromolecules* **2,** 241 (1969).
71. Y. Inoue, R. Chûjô, and A. Nishioka, *J. Polym. Sci. Polym. Phys. Ed.* **11,** 393 (1973).
72. Y. Inoue, R. Chûjô, A. Nishioka, S. Nozakura, and H. Iimuro, *Polym. J.* **4,** 244 (1973).
73. B. Ibrahim, A. R. Katritzky, A. Smith, and D. E. Weiss, *J. Chem. Soc., Perk. Trans.* **II,** 1537 (1974).
74. I. R. Peat and W. F. Reynolds, *Tetrahedron Lett.* No. 14, 1359 (1972).
75. B. D. Coleman and T. G. Fox, *J. Polym. Sci. Polym. Chem. Ed.* **1,** 3183 (1963).
76. K. F. Elgert, R. Wicke, B. Stützel, and W. Ritter, *Polymer* **16,** 465 (1975).
77. Y. Inoue, A. Nishioka, and R. Chûjô, *Makromol. Chem.* **156,** 207 (1972).
78. S. Fujishige and I. Ando, *Makromol. Chem.* **177,** 2195 (1976).
79. H. Girad and P. Monjol, *C. R. Hebd. Seances Acad. Sci., Ser. C,* **279**(13), 553 (1974).
80. K. Matsuzaki, T. Kanai, T. Kawamura, S. Matsumoto, and T. Uryu, *J. Polym. Sci. Polym. Chem. Ed.* **11,** 961 (1973).
81. K. Matsuzaki, H. Ito, T. Kawamura, and T. Uryu, *J. Polym. Sci. Polym. Chem. Ed.* **11,** 971 (1973).
82. Y. Inoue, K. Koyama, R. Chûjô, and A. Nishioka, *Makromol. Chem.* **175,** 277 (1974).
83. J. Schaefer, *Macromolecules* **4,** 105 (1971).
84. D. K. Dalling and D. M. Grant, *J. Amer. Chem. Soc.* **89,** 6612 (1967).
85. A. Zambelli, G. Gatti, C. Sacchi, W. O. Crain, Jr., and J. D. Roberts, *Macromolecules* **4,** 475 (1971).
86. C. E. Wilkes, C. J. Carman, and R. A. Harrington, *J. Polym. Sci., Symp. No. 43,* 237 (1973).
87. C. J. Carman and K. C. Baranwal, *Rub. Chem. Tech.* **48,** 705 (1975).
88. Y. Tanaka and K. Hatada, *J. Polym. Sci. Polym. Chem. Ed.* **11,** 2057 (1973).
89. G. J. Ray, P. E. Johnson, and J. R. Knox, *Macromolecules* **10,** 773 (1977).
90. D. E. Dorman, E. P. Otocka, and F. A. Bovey, *Macromolecules* **5,** 574 (1972).
91. M. Delfini, A. Segre, and F. Conti, *Macromolecules* **6,** 456 (1973).
92. J. Schaefer, *Macromolecules* **4,** 107 (1971).
93. A. R. Katritzky and D. E. Weiss, *J. Chem. Soc., Perk. Trans.* **II,** 21 (1975).
94. A. R. Katritzky and D. E. Weiss, *J. Chem. Soc., Perk. Trans.* **II,** 27 (1975).
95. D. E. Dorman, M. Jautelat, and J. D. Roberts, *J. Org. Chem.* **36,** 2757 (1971).
96. A. R. Katritzky and D. E. Weiss, *Chem. Brit.* **12,** 45 (1975).
97. A. R. Katritzky and D. E. Weiss, *J. Chem. Soc., Perk. Trans.* **II,** 1542 (1974).
98. A. R. Katritzky, A. Smith, and D. E. Weiss, *J. Chem. Soc., Perk. Trans.* **II,** 1547 (1974).

Index

Q

R

S

A
B 7
C 8
D 9
E 0
F 1
G 2
H 3
I 4
J 5